Beast ACADEMY PUZZLES

LEVEL 4

AGES 10 to ∞

By Art of Problem Solving

Published by:
AoPS Incorporated
15330 Avenue of Science
San Diego, CA 92128
info@BeastAcademy.com

ISBN: 978-1-934124-59-8

Beast Academy is a registered trademark
of AoPS Incorporated.

Written by Chris Page, Anderson Wang,
Palmer Mebane, and Jason Batterson
Book Design by Doğa Arı
Illustrations by Erich Owen

Visit the Beast Academy website at
BeastAcademy.com.

Visit the Art of Problem Solving website at
artofproblemsolving.com.

Printed in the United States of America.
First Printing 2022.

CONTENTS

ABOUT THIS BOOK

Beast Academy Puzzles 4 contains more than 500 puzzles in 12 different styles. Every puzzle style is aligned with the broader Beast Academy level 4 math curriculum. Whether used on their own or as part of the complete Beast Academy curriculum, these puzzles will delight and entertain puzzle solvers of all ages.

The puzzles in this book cover a broad range of topics including shape classification, symmetry, multiplication, exponents, factoring, logic, fractions, integers, and decimals as taught in the Beast Academy level 4 series. The difficulty ranges from straightforward puzzles meant to give a feel for how each puzzle works to diabolical stumpers written by world puzzle champion Palmer Mebane.

WHY PUZZLES?

Entertainment

Puzzles intrigue us and capture our attention in ways that many other problems don't. What makes puzzles so captivating?

- **Breakthroughs.** The "Aha!" moments of ingenuity and insight that come when solving a well-written puzzle are energizing.

- **Satisfaction.** Not every puzzle has an "Aha!" moment of inspiration. Many involve a series of steps that are satisfying and encouraging in their own way.

- **Accomplishment.** Solving a puzzle that is just at the edge of your ability level gives a wonderful sense of achievement.

- **Gratification.** Unlike many other problems you face, it's often immediately obvious when you've solved a puzzle correctly.

Enrichment

Solving puzzles makes us smarter. What do we learn?

- **Problem Solving.** The skills we learn by solving puzzles—observing, testing, fiddling, and making connections—help us become better, more resilient problem solvers in other areas.

- **Math Skills.** Every puzzle in this book was written to reinforce specific math skills. Puzzles take the monotony out of skill drill and make practice fun.

- **Spatial Reasoning.** Many puzzles require elements of path tracing or grouping that help us build spatial awareness.

- **Pattern Recognition.** Solving puzzles helps us recognize patterns and encourages us to search for new ones.

Dot Puzzles

Spiral Galaxies

Product Placement

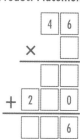

Pyramid Descent

Dutch Loop

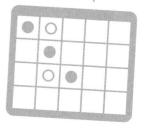

Hive

Factor Cave

Factor Blobs

Fraction Sumdoku

Sum Squares

Paint the Town

Decimal Numbercross

USING THIS BOOK

This book is divided by puzzle type into 12 sections, followed by HINTS beginning on page 165 and SOLUTIONS beginning on page 191.

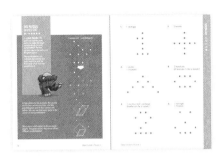

Each section includes instructions, a solved example, and difficulty ratings. The ratings at the edge of the page refer to the difficulty of the hardest problem on the page and are highly subjective.

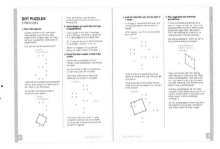

At the end of each puzzle set is a STRATEGIES section. We recommend reading the strategies section even if you've already solved all of the puzzles. There may be an approach you haven't considered.

Supplementing the BA Curriculum

If you are using this book to supplement the Beast Academy math curriculum, below is a list of the different puzzle types, the chapters they supplement in BA level 4, and what math skills they reinforce.

Dot Puzzles	Chapter 1	Shape identification and classification. New Puzzle!
Spiral Galaxies	Chapter 1	Symmetry. Find more like this in BA Online.
Product Placement	Chapter 2	Multiplication. Find more like this in Practice book 4A.
Pyramid Descent	Chapters 3&7	Exponents and factoring. Find more in Practice books 4A and 4C.
Dutch Loop	Chapter 6	Logic and spatial reasoning. Classic puzzle.
Hive	Chapter 6	Logic. Find more like this in Practice book 4B.
Factor Cave	Chapter 7	Factoring. New puzzle!
Factor Blobs	Chapter 7	Factoring. Find more in Practice book 4C.
Fraction Sumdoku	Chapter 8	Adding fractions. New puzzle!
Sum Squares	Chapter 9	Adding integers. Find more in Practice book 4C.
Paint the Town	Chapter 10	Fractions. Adaptation of Fraction Fill in Practice book 4D.
Decimal Numbercross	Chapter 11	Converting fractions to decimals. New puzzle!

DOT PUZZLES
DIFFICULTY LEVEL:

★—★★★★★

In a **Dot Puzzle**, the goal is to use the given points to make the target quadrilaterals. In each quadrilateral, the four corners need to be on four of the given points.

The quadrilaterals can intersect and overlap, but each point can only be the corner of at most one quadrilateral.

1 square and 1 parallelogram

In the solution to the example, the square and its four corners are blue, and the parallelogram and its four corners are yellow. Every point is the corner of at most one quadrilateral!

This is not a valid solution to the example puzzle. The green point is the corner of two different quadrilaterals!

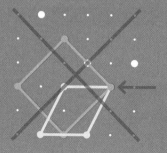

1. 1 rectangle

2. 2 squares

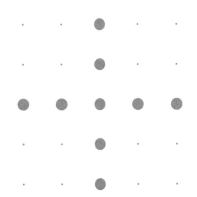

3. 1 square
1 trapezoid

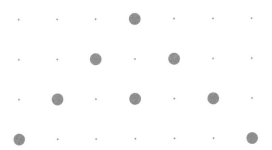

4. 2 rhombuses
(At least one is also a square.)

5. 1 rhombus and 1 rectangle
(Neither can be a square.)

6. 1 rectangle
1 rhombus

7. 1 square

8. 2 rhombuses
(Neither can be a square.)

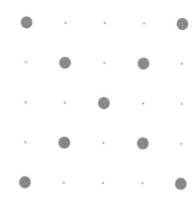

9. 1 square

10. 1 parallelogram

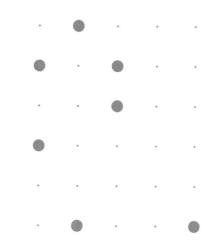

11. 1 rectangle

12. 1 square
1 parallelogram

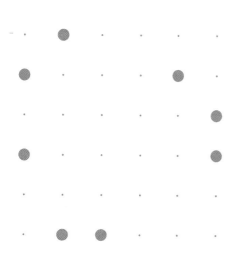

13. 1 rhombus

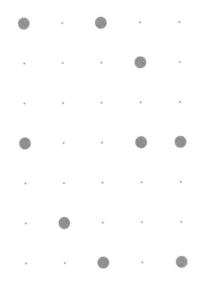

14. 2 parallelograms

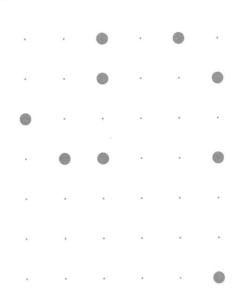

15. 2 rectangles

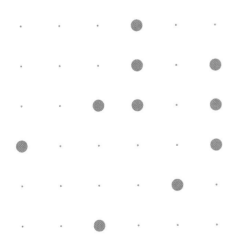

16. 1 square
1 rectangle

Hints on pages 166-167

17. 2 squares
 H 1 rectangle

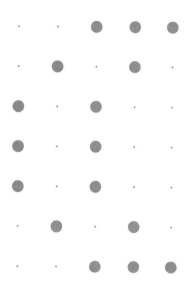

18. 2 rectangles
 H

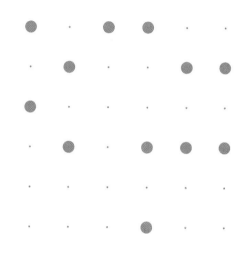

19. 1 square
 H 1 parallelogram

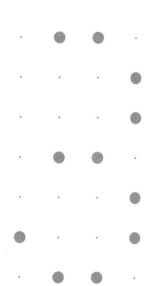

20. 2 rectangles
 H (Neither can be a square.)

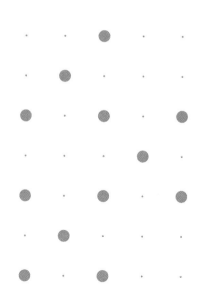

21. 3 squares

22. 3 squares

23. 3 rectangles
(At least one is also a square.)

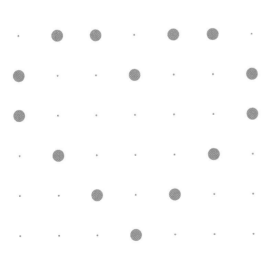

24. 2 rhombuses

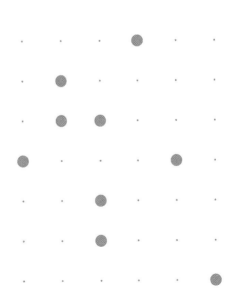

H Hints on pages 166-167

25. Ⓗ 1 rhombus and 1 rectangle
(Neither can be a square.)

26. Ⓗ 2 rectangles

27. Ⓗ 3 squares

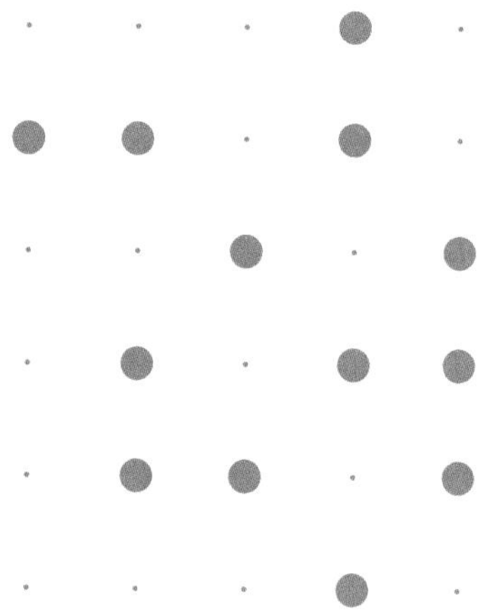

28. Ⓗ 3 squares

12 Ⓗ Hints on pages 166-167

29. Ⓗ 2 parallelograms
(Neither can be a rectangle.)

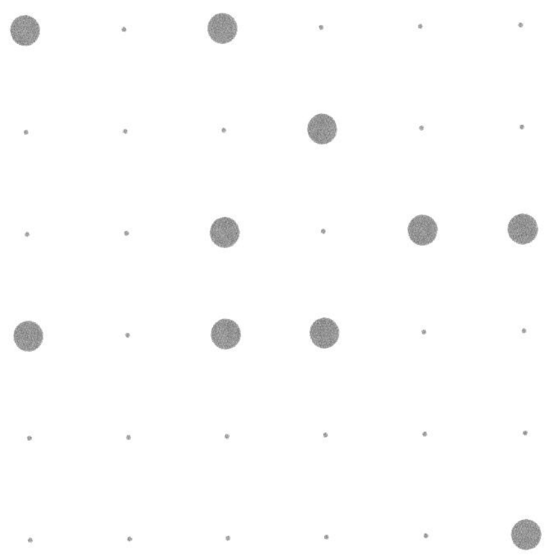

30. Ⓗ 1 square
1 rectangle
1 parallelogram

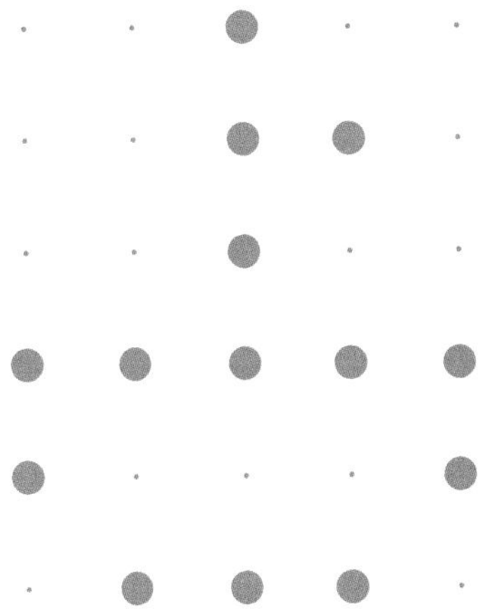

31. Ⓗ 4 squares

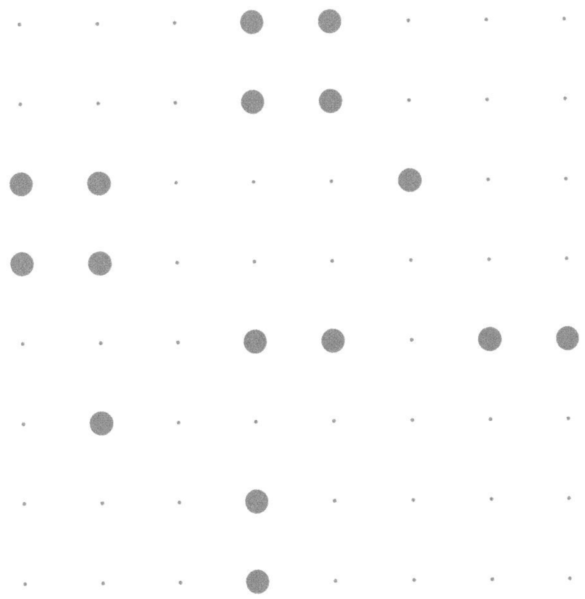

32. Ⓗ 4 rectangles
(At least one is also a square.)

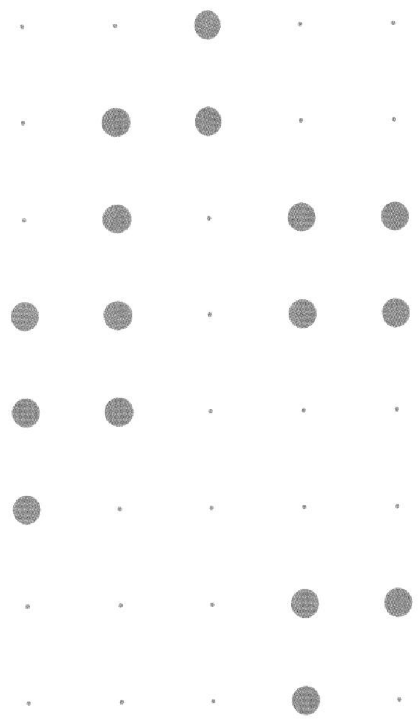

Ⓗ Hints on pages 166-167

33.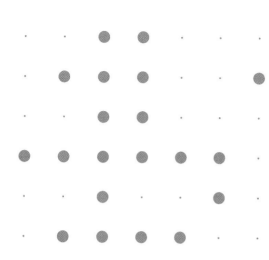
H 1 square, 1 rhombus,
1 rectangle, and 1 parallelogram

34. 5 squares
H

35. 4 rhombuses (At least one is also a square.)
H

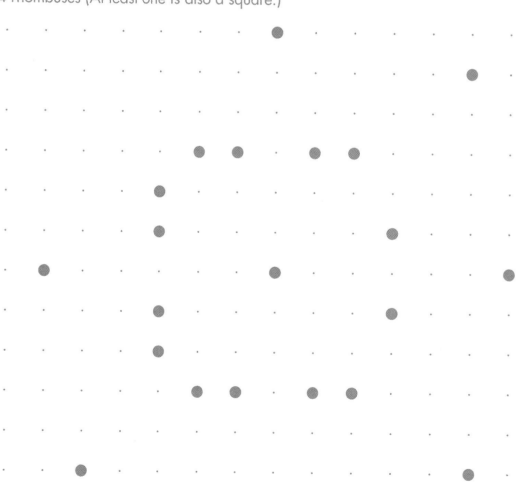

H Hints on pages 166-167

36. 2 squares
2 parallelograms

37. 3 rectangles and 1 rhombus
(None are squares.)

38. 3 rectangles (At least one is also a square.)

Hints on pages 166-167

DOT PUZZLES
STRATEGIES

1. Start with squares.

Squares have the most "restrictions" because they must have four right angles and four equal sides. So, there are fewer squares than other types of quadrilaterals.

How can we start the puzzle below?

There are many different parallelograms we can draw using four of these points. But, there is only one square.

We connect four points to make a square as shown below.

Then, we look for a way to make a parallelogram that does not use any of these points.

2. Some shapes are more than one type of quadrilateral.

Every square is also both a rectangle and a rhombus, since every square has four right angles and four equal sides.

So, if a puzzle asks us to find rhombuses or rectangles, squares count, too!

When this happens, the puzzle will always include a note as a reminder.

3. Count the total number of dots in the puzzle.

Since every quadrilateral has four corners, every quadrilateral uses exactly four dots.

So, the number of dots in a puzzle tells us how many dots are not used.

How many of the dots in the puzzle below will not be part of a square?

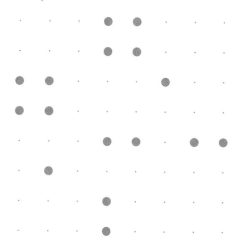

We count a total of 16 dots. To draw 4 squares, we must use 4×4=16 dots. So, every dot in the puzzle must be the corner of a square!

4. Look for dots that can only be part of 1 shape.

In strategy 3, we learned that every dot in the puzzle below must be part of a square.

What square(s) can the blue dot below be a part of?

4 squares

There is only one square that can be drawn on the grid that uses the blue dot as a corner.

Since every dot in this puzzle must be used, we know this square must be part of the solution.

4 squares

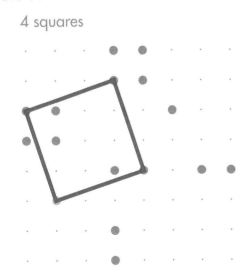

5. Stay organized and eliminate possibilities.

If there are limited possibilities for a point or shape, try them all. Try to rule out possibilities that don't work to avoid trying them again later. (Be careful! Don't rule anything out unless you're sure you haven't missed anything.)

Can this parallelogram be part of the solution to the puzzle below?

2 parallelograms
(Neither can be a rectangle.)

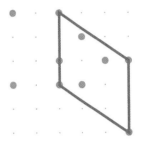

There are 6 points left. We carefully check groups of 4 points to see if they make a parallelogram. One way to do this is by selecting two points to leave out, then checking if the remaining four points make a parallelogram.

The only parallelogram we can make using four of the remaining six points is a square, but this doesn't count because every square is also a rectangle (strategy 2).

So, this parallelogram cannot be part of the solution to this puzzle. We erase it and make a note not to try it again.

6. Check for parallel sides.

We can find out if two sides are parallel by counting "hops" between dots on their sides.

Sides that are parallel have the same pattern of hops in the same directions from one dot to the next.

Is the shape below a parallelogram?

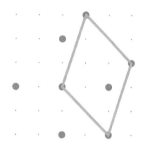

The shape *looks* like a parallelogram, but how can we tell? To find out, we count how many hops we need to get from each corner to the next.

For the sides below, we hop 2 dots right and 2 dots down to get from one corner to the other. So, they are parallel.

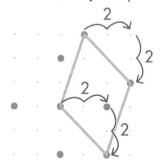

For the other two sides, we hop 3 dots up and 1 dot right to get from one corner to the other. So, they are also parallel.

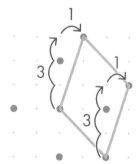

Since opposite sides are parallel, this quadrilateral is a parallelogram.

7. Check for equal side lengths.

In all of the dot puzzles in this book, we can find out if two lengths are equal by counting hops along their sides.

Sides that are the same length have the same pattern of hops from one end to the other, but not always in the same order or direction. For example, hopping 3 dots right and 2 dots up is the same distance as hopping 2 dots left and 3 dots down.

Is the green quadrilateral below a rhombus?

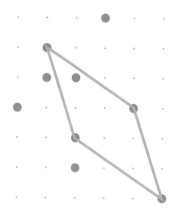

The shape *looks* like a rhombus, but how can we tell for sure? To compare the lengths of its sides, we count how many hops it takes to get from one corner to the next.

On two of the sides, we hop 3 dots one way and 1 dot another way to get from one corner to the other.

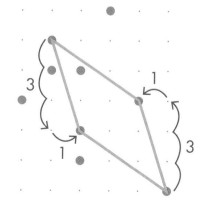

But for the other two sides, we hop 3 dots one way and <u>2</u> dots another way to get from one corner to the other.

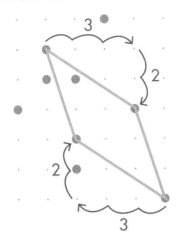

So, these sides are a little longer than the other two. Since the four sides are not all the same length, this shape is not a rhombus.

(We will learn how to find these lengths when we learn about the Pythagorean Theorem in Beast Academy 5D.)

8. Check for squares.

We can count hops along sides to check squares, too.

Find all of the squares that have a corner on the blue dot below.

There are two squares we can make that have a corner on the blue dot.

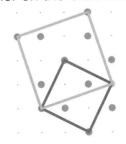

The pattern of hops along the sides of a square will always be the same number of hops, but rotated as we move around the square.

For example, from the top-left corner to the top-right corner of the green square below, we hop up 1, and right 3.

If we turn the grid on its side, we make the same hops to get to the next corner.

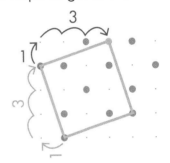

We can continue around the square.

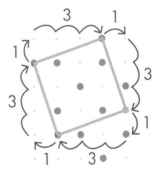

Similarly, we can find the pattern of hops for the smaller square below.

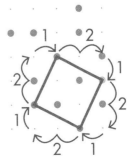

9. Check for right angles.

We can even count hops to see if two sides meet at a right angle.

Is either quadrilateral below a rectangle?

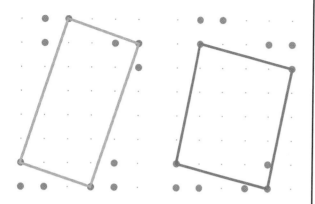

We could use the corner of a piece of paper to see if the angles are right, but these angles are all really close!

So, we look at the dots in the grid and use what we know about squares from strategy 8.

For the top side of the green quadrilateral, we go right 3 hops and down 1 hop to get from one corner to the other.

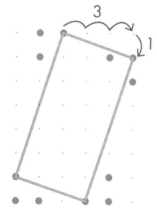

If we turn this pattern around the corner, going down 3 and left 1, we reach a point on the side of the rectangle.

So, these dots make the corner of a square, and the top-right corner must be a right angle!

We can continue around the shape to see that the green quadrilateral can be made from two squares. So, it is a rectangle.

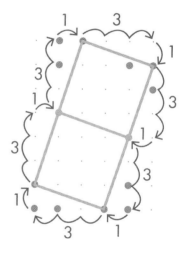

For the top side of the purple quadrilateral, we go right 4 hops and down 1 hop to get from one corner to the other.

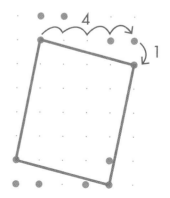

To make a right angle, we turn this pattern around the corner by going down 4 and left 1.

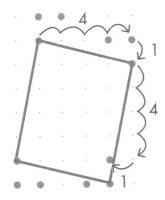

But, this point is not on a side of the purple quadrilateral.

From one corner to the other of a long side, we go down 5 and left 1. So, the top-right angle is a little bigger than a right angle.

We've drawn a square in the diagram below to show what a right angle would look like.

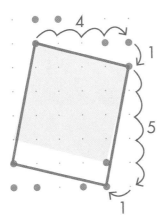

Since the purple quadrilateral does not have four right angles, it is not a rectangle.

10. Check for rhombuses.

Rhombuses have special properties that can help us find them. In a rhombus, the diagonals cross at a right angle. And, the diagonals split each other in half.

So, when looking for rhombuses, it often helps to look for diagonals like the pairs below.

11. Look for one pair of sides.

In any parallelogram, if one pair of opposite sides are parallel and the same length, then the other pair of sides will also be parallel and the same length.

So, when looking for parallelograms, we can look for just one pair of opposite sides. Look for pairs of dots that are separated by the same numbers of hops like the ones shown below.

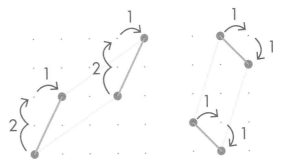

SPIRAL GALAXIES
DIFFICULTY LEVEL:
★—★★★★★

In a **Spiral Galaxies** puzzle, the goal is to draw walls that divide the grid into "galaxies" that have rotational symmetry.

Every galaxy must contain a galaxy symbol  at its center.

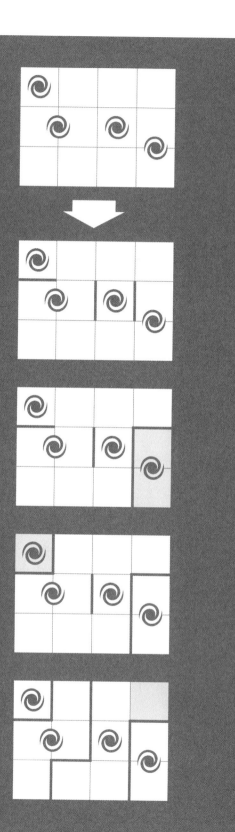

Since each galaxy can only have one symbol at its center, we draw walls between squares that contain symbols.

Next, the two highlighted squares must be part of the galaxy with the symbol between them. We can't add more squares to this galaxy and keep its rotational symmetry.

So, we draw walls around these two squares.

Similarly, we cannot add more squares to the galaxy in the top-left corner, so we draw walls to enclose it.

Finally, the square in the top-right corner can only be part of the galaxy that has its center in the third column. We enclose the galaxy as shown.

Every galaxy has rotational symmetry, so we are done!

Beast Academy Puzzles 4

☆
☆ ☆
★ ☆

1.
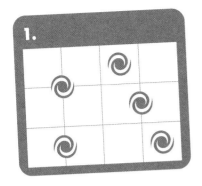

2.
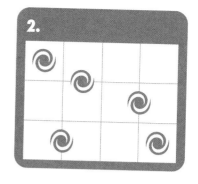

3.
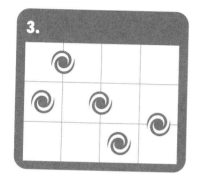

4.
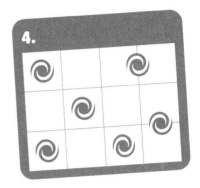

5.

6.
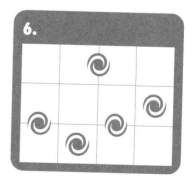

7.

8.
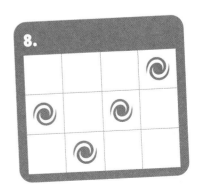

★ ★ ☆ ☆ ☆

15.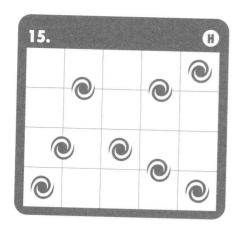

16.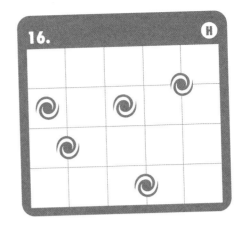

17.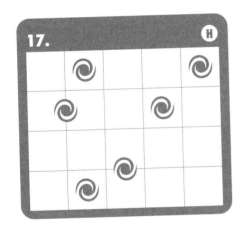

18.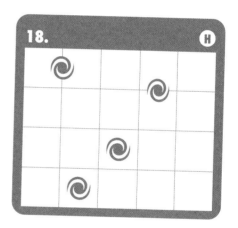

19.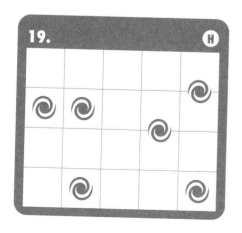

20.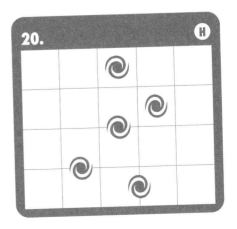

H Hints on pages 168-169

☆
☆
★
★
★

21.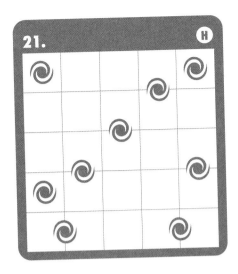

22.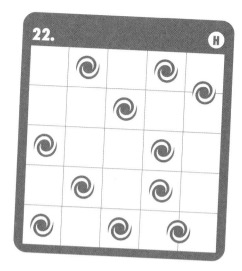

23.

24.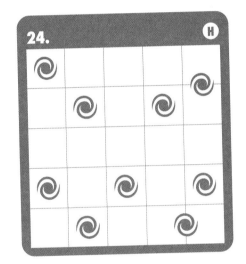

25.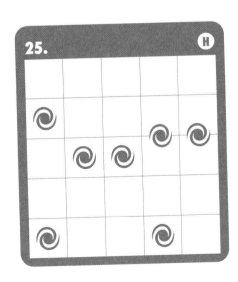

26.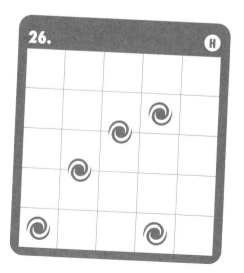

27.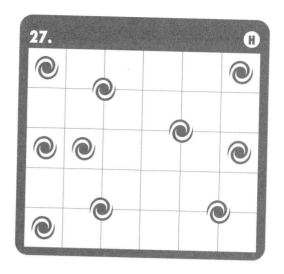

28.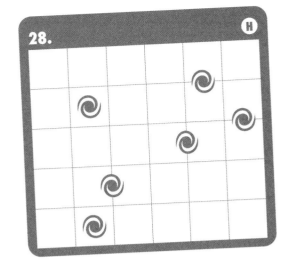

29.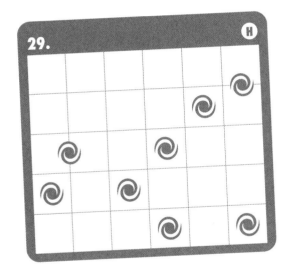

30.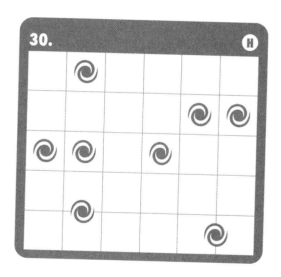

31.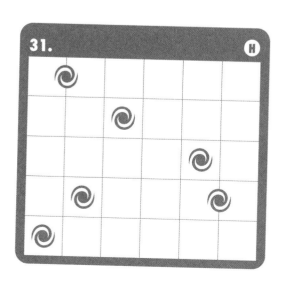

32.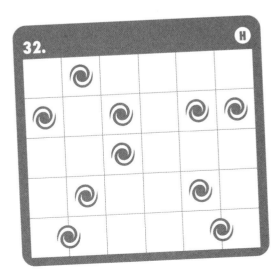

Hints on pages 168-169

☆
★
★
★
★

33. H

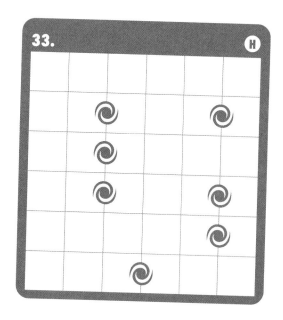

34. H

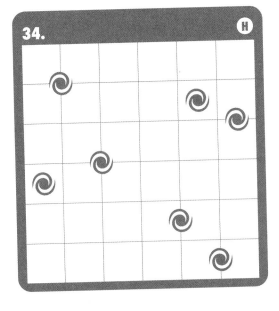

35. H

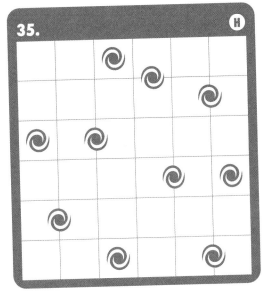

36. H

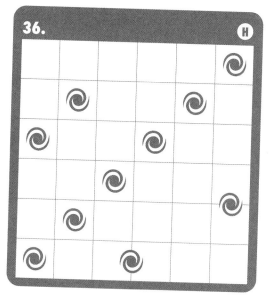

37. H

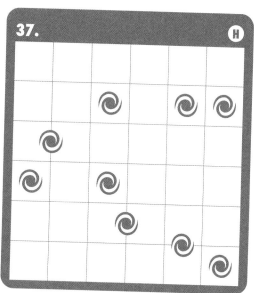

38. H

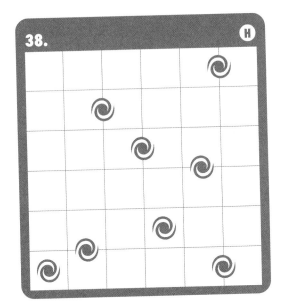

39.

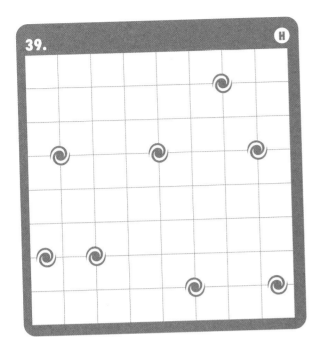

40.

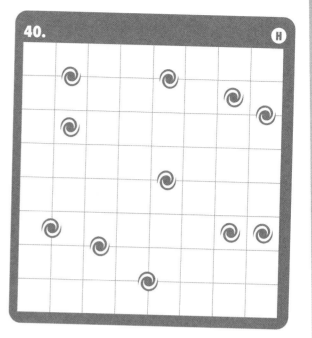

41.

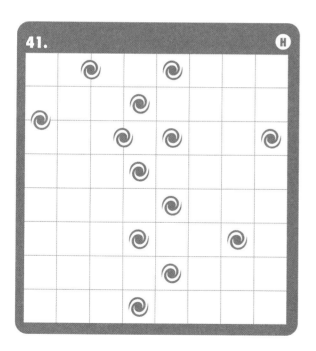

42.

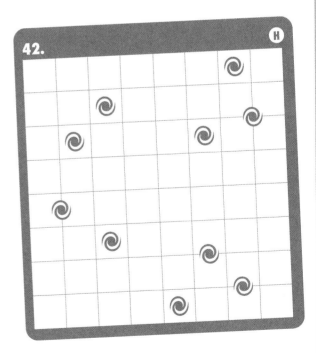

Hints on pages 168-169

SPIRAL GALAXIES
STRATEGIES

1. Draw walls between squares that cannot be part of the same galaxy.

If two squares that are next to each other both contain galaxy centers, then they must be part of different galaxies.

We can draw walls between squares that are in different galaxies.

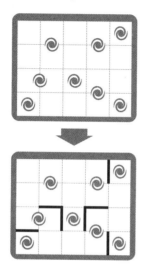

2. Squares in a galaxy must have rotational symmetry.

If a square is part of a galaxy, then its rotation halfway around the center of the galaxy is also part of the galaxy.

Can the blue square below be part of the yellow galaxy?

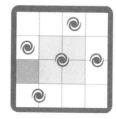

For the blue square to be part of the yellow galaxy, its rotation halfway around

the center of the galaxy must also be part of the galaxy.

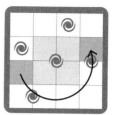

Since this square is part of a different galaxy, neither blue square can be part of the yellow galaxy.

We draw walls to show that these squares are not part of the yellow galaxy.

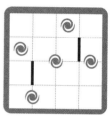

3. Draw walls around squares that touch the edges.

Can the blue square be part of the yellow galaxy below?

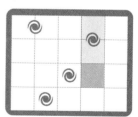

For the blue square to be part of the yellow galaxy, its rotation halfway around the center of the galaxy must also be part of the galaxy.

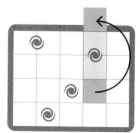

Since the rotation is outside the grid, the blue square cannot be in the galaxy.

We draw a wall to mark the edge of the galaxy.

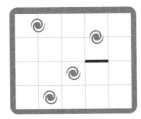

4. Draw walls around galaxies in corners.

Which other squares can be part of the yellow galaxy below?

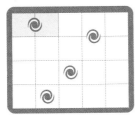

We can use strategy 3 to show that there cannot be any squares below or to the right of this galaxy. So, we draw walls that surround the galaxy as shown below.

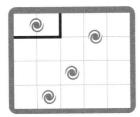

5. Walls that border a galaxy have rotational symmetry.

In the puzzle below, we've used earlier strategies to draw some walls around galaxies. What additional wall can we draw around the yellow galaxy?

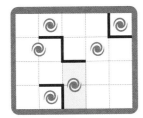

If we know a wall borders a galaxy, then

we can also draw its rotation halfway around the galaxy.

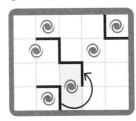

In general, whenever we draw a wall that borders a galaxy, we can also draw its rotation halfway around the center of the galaxy.

6. Connect squares that must be part of the same galaxy.

If we are sure that squares are part of the same galaxy, we can connect them to help us keep track.

In the puzzle below, we've used strategy 4 to outline galaxies in the corners. Which squares must be part of the galaxy with its center highlighted below?

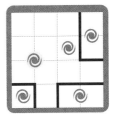

First, the four squares that contain the galaxy's center must be part of the galaxy. We connect these squares.

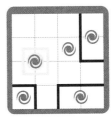

Then, the square in the bottom row can only be connected to this galaxy. To keep the galaxy symmetrical, the square in the top-left corner must also be part of this galaxy.

We draw connections to these squares.

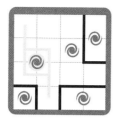

7. Find squares that can only be part of one galaxy.

Which galaxy symbol is the center of the galaxy that contains the yellow square below?

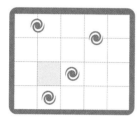

The square cannot be in the same galaxy as any of the galaxy centers below, because its rotation would be outside the grid.

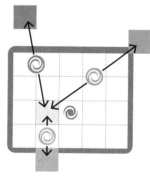

So, the square must be part of the galaxy with the center to its right. Its rotation around the center must also be part of the same galaxy.

We draw a line to highlight that these three squares are connected.

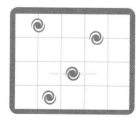

PRODUCT PLACEMENT

DIFFICULTY LEVEL:

In a **Product Placement** puzzle, the goal is to fill each empty box with a digit to complete the product.

The multiplication algorithm we use in these puzzles uses partial products and may be different from others you have seen.

All of the puzzles in this book multiply a number with two or more digits by a one-digit number.

We multiply the one-digit number by each place value in the larger number, then add the partial products as shown below.

Ex.
```
   2 3 4
 ×     5
```

Step 1:   2 3 **4**
5×4   ×   5
          2 0

Step 2:   2 **3** 4
5×30   ×   5
          2 0
        1 5 0

Step 3:   **2** 3 4
5×200   ×   5
          2 0
        1 5 0
      1 0 0 0

Step 4:   2 3 4
Add   ×   5
          2 0
        1 5 0
      +1 0 0 0
      1 1 7 0

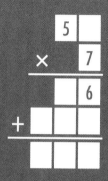

The first partial product gives us 7×■=■6.

The only digit we can multiply by 7 to get a number that ends in 6 is 8.

So, we have 7×**8**=**5**6.

The second partial product in this puzzle is 7×50=350.

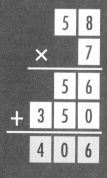

Finally, we add the two partial products to fill in the final product.

56+350=406.

★ ☆ ☆ ☆ ☆

1.

```
    8   5
×       □
─────────
    1   0
+ □ □ □
─────────
  □ □ □
```

2.

```
    2   8
×       □
─────────
    □ □
+ 1 6 0
─────────
  □ □ □
```

3.

```
    7   9
×       □
─────────
    3   6
+ □ □ □
─────────
  □ □ □
```

4.

```
    6   □
×       4
─────────
    1   6
+ □ □ □
─────────
  □ □ □
```

5.

```
    □   6
×       7
─────────
    □ □
+ 1 4 0
─────────
  □ □ □
```

6.

```
    7   □
×       7
─────────
    6   3
+ □ □ □
─────────
  □ □ □
```

7.

$$
\begin{array}{r}
8\ \square \\
\times\ \ \ 6 \\
\hline
\square\ \square \\
+\ \square\ \square\ \square \\
\hline
5\ 1\ 0
\end{array}
$$

8.

$$
\begin{array}{r}
5\ \square \\
\times\ \ \ 7 \\
\hline
\square\ \square \\
+\ \square\ \square\ \square \\
\hline
3\ 7\ 8
\end{array}
$$

9.

$$
\begin{array}{r}
3\ \square \\
\times\ \ \ 6 \\
\hline
\square\ \square \\
+\ \square\ \square\ \square \\
\hline
2\ 2\ 2
\end{array}
$$

10.

$$
\begin{array}{r}
\square\ \square \\
\times\ \ \ 7 \\
\hline
2\ 8 \\
+\ \square\ \square\ \square \\
\hline
1\ 6\ 8
\end{array}
$$

11.

$$
\begin{array}{r}
9\ \square \\
\times\ \ \ \square \\
\hline
\square\ \square \\
+\ 5\ 4\ 0 \\
\hline
5\ 8\ 8
\end{array}
$$

12.

$$
\begin{array}{r}
\square\ 7 \\
\times\ \ \ \square \\
\hline
\square\ \square \\
+\ 1\ 5\ 0 \\
\hline
1\ 7\ 1
\end{array}
$$

13.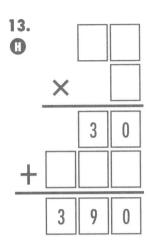

14.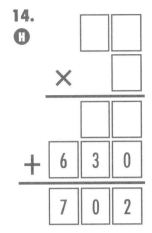

15.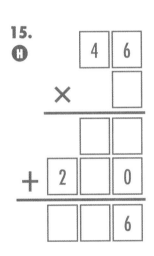

16.

17.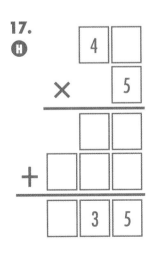

18.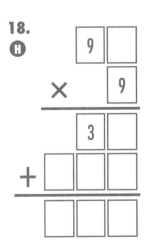

Hints on pages 170-171

19.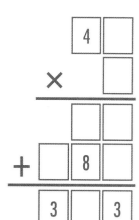

```
        4 ☐
      ×   ☐
      ─────
        ☐ ☐
  +   ☐ 8 ☐
  ───────────
    3 ☐ 3
```

20.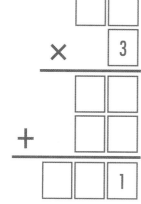

```
        ☐ ☐
      ×   3
      ─────
        ☐ ☐
  +   ☐ ☐
  ─────────
    ☐ ☐ 1
```

21.

```
      ☐ ☐ 3
    ×     ☐
    ───────
        ☐ 1
      ☐ 2 ☐
  + ☐ 3 ☐ ☐
  ─────────
    ☐ ☐ ☐ ☐
```

22.

```
      ☐ ☐ ☐
    ×     ☐
    ───────
        ☐ 0
      ☐ 2 ☐
  + ☐ 8 ☐ ☐
  ─────────
    3 ☐ 4 ☐
```

23.

```
      ☐ ☐ ☐
    ×     ☐
    ───────
        4 9
        ☐ ☐
  +   ☐ ☐ ☐
  ───────────
    8 ☐ ☐
```

24.

```
      ☐ ☐ 6
    ×     6
    ───────
        ☐ ☐
        ☐ ☐
  + 4 ☐ ☐ ☐
  ─────────
    ☐ 4 1 ☐
```

25.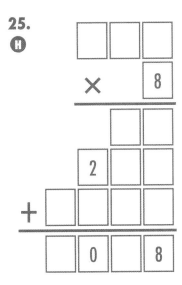

26.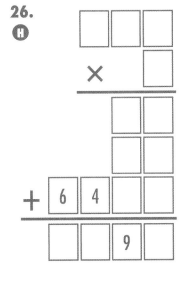

27.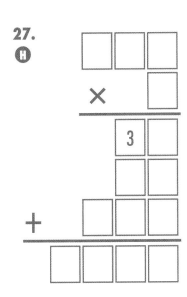

28.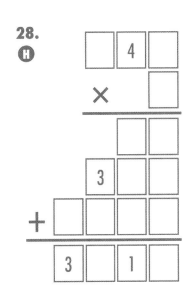

29.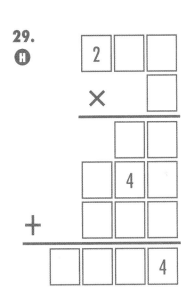

30.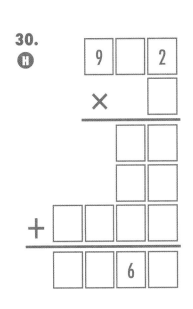

H Hints on pages 170-171

31.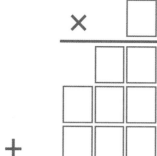

32.

33.

34.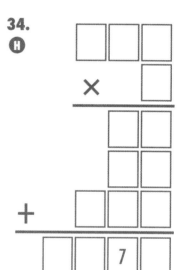

35.

36.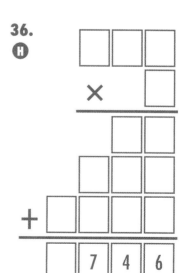

Hints on pages 170-171

37.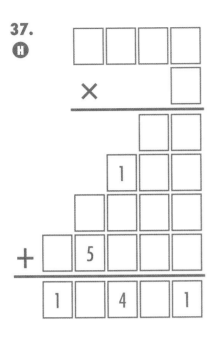

38.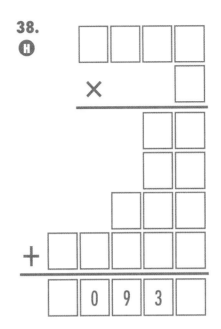

39.

40.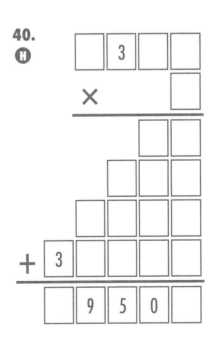

PRODUCT PLACEMENT
STRATEGIES

1. Place trailing 0's in the partial product(s).

Many of the rightmost digits in the partial products must be 0's. We can write these in to begin each puzzle.

Where can we place 0's in the partial products below?

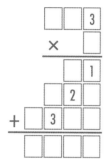

The second partial product is a multiple of ten, so it ends in 0.

The third partial product is a multiple of one hundred, so it ends in two 0's.

We write these 0's in as shown.

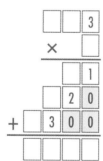

2. After writing 0's in the partial products, ignore them!

Focus on the digits that come before the trailing zeros in the partial products.

This helps us think of each partial product as a number less than 100 and makes finding factors easier.

3. Find partial products we can compute.

What is the second partial product?

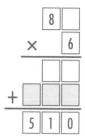

We can simply multiply 6×80=480.

Or, we can use strategies 1 and 2.

First, we place the trailing 0. Then, we ignore the trailing 0 and multiply 6×8=48 to fill in the remaining digits.

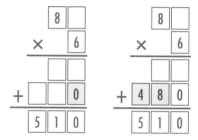

4. Find factors (or digits of factors) that we can compute.

What digit fills the highlighted box?

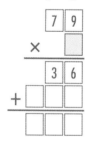

The first partial product gives us $\boxed{}$×9=36.

Since $\boxed{4}$×9=36, we place a 4 in the box.

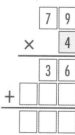

5. Find the missing number in a sum.

What is the highlighted partial product?

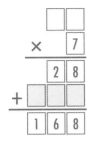

We add the partial products to get the total product. We have 28+☐☐☐=168.

So, the second partial product is 168−28=$\boxed{1}\boxed{4}\boxed{0}$.

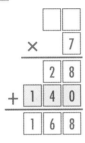

6. Find missing digits using the sums of partial products.

We can use known digits in the columns of partial products to find missing digits.

What digit fills the highlighted box?

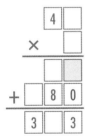

In the ones column of the sum, we have ☐+0=3, so the ones digit of the first partial product is 3.

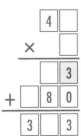

When using this strategy, remember that sometimes there are tens or hundreds or thousands regrouped from another column that will affect the sum.

7. Look for partial products that have unique or limited factors.

Some partial products have limited factor pairs that can be used.

What digits fill the boxes below?

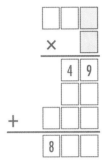

The only way to get a partial product of 49 is with 7×7=49. So, we place 7's as shown.

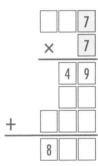

8. Look for partial products that have unique digits.

What digit fills the highlighted box?

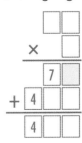

The only product of two digits from 70 to 79 is 8×9=72.

So, the missing digit is a 2.

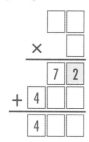

9. Find the only factor that gives the correct digit of a partial product.

What digits fill the highlighted boxes?

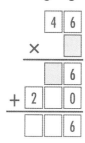

The first partial product is ☐×6=☐6. So, ☐6 is a multiple of 6.

The only two-digit multiple of 6 that ends in 6 is 36. So, the missing factor is 6 and the first partial product is 36.

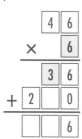

10. Find common factors of partial products.

What digit fills the highlighted box?

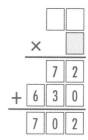

For the first partial product, only 9×8=72.

For the second partial product, only 9×7=63 (using strategy 2, we ignore the trailing zero).

So, the highlighted digit can only be a 9.

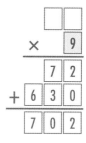

11. Pay attention to the numbers of digits in the partial and final products.

The numbers of digits in the products can give us clues about what those digits are.

What digits fill the highlighted boxes?

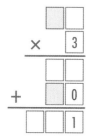

Ignoring the trailing zero, the second partial product is 3×☐=☐.

So, we have either 3×1=3, 3×2=6, or 3×3=9.

This means that the second partial product is either 30, 60, or 90.

Since the final product has three digits, and the first partial product is at most 3×9=27, the second partial product cannot be 30 or 60.

So, we must use 3×3=9.

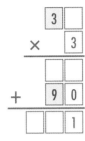

PYRAMID DESCENT
DIFFICULTY LEVEL:

★ — ★ ★ ★ ★

In a **Pyramid Descent** puzzle, the goal is to find a path of touching squares, one per row, from the top to the bottom of the pyramid.

The product of the numbers on the path must equal the number shown above the pyramid.

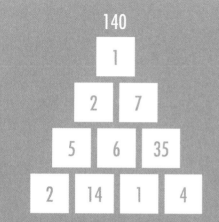

We use the prime factorization of 140 to help us find the path.

140=2×2×5×7. So, our path must include two 2's, one 5, and one 7.

We factor the composite numbers in the pyramid to help us find these factors.

6=2×3, 35=5×7, 14=2×7, and 4=2×2.

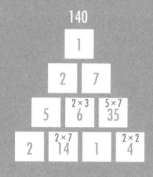

140 does not have any 3's in its prime factorization. So, we cross out the 6.

Then, if we include the 7 in the second row, the path must also include the 5×7=35 in the third row. Since 140 only has one 7 in its prime factorization, we cannot include two 7's in the path, and we cross them out.

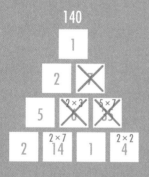

There is only one remaining factor of 5 and only one remaining factor of 7. We draw the only possible path that includes those two numbers.

We check our path. 1×2×5×14 has two 2's, one 5, and one 7. So, 1×2×5×14=140.

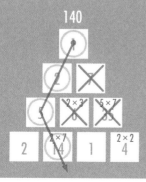

1.

12

2

3 1

1 2 3

2.

27

3

3^2 3

3 3 3^2

3.

65

1

3 5

19 11 13

4.

5^4

5

25 5

125 25 5

5.

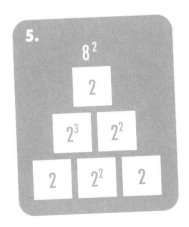

8^2

2

2^3 2^2

2 2^2 2

6.

108

6

3 4

6 2 9

7.

3^3

	1		
	1	3	
3	3	1	
3	1	3	3

8.

300

	2		
	5	10	
10	3	10	
2	5	2	5

9.

315

	5		
	3	7	
7	6	14	
10	3	21	15

10.

11^7

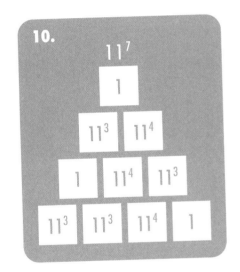

	1		
	11^3	11^4	
1	11^4	11^3	
11^3	11^3	11^4	1

11.

14^2

	2		
	2	7	
2	2	2	
7	7	2	2

12.

224

	4		
	7	4	
2	7	4	
1	2	7	4

13. 15^2

14. 2^{30}

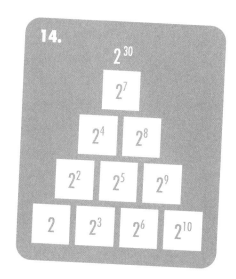

15. 48

16. 12^3

17. 18^2

18. 910

19. 7^{13}

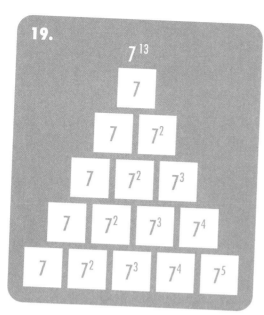

20. 270

21. 192

22. 8^7

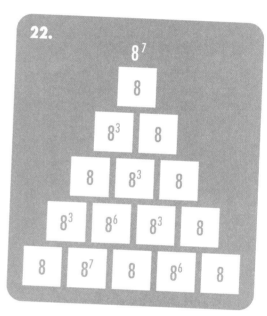

23. 2100

24. 4^5

25.

960

26.

6^8

27.

1470

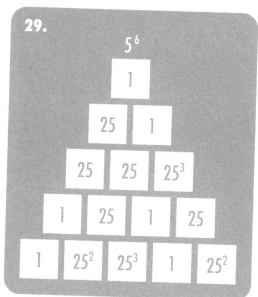

28.

21^6

29.

5^6

30.

1440

31.

3025

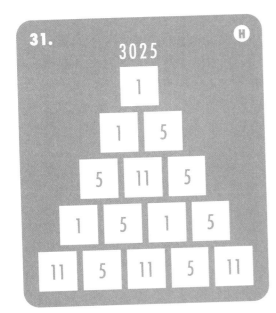

32.

30^4

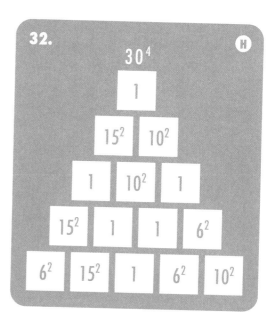

33.

4864

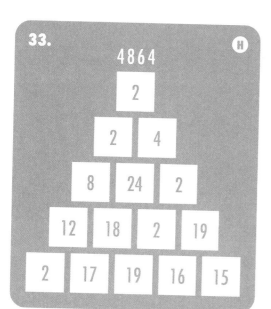

34.

1056

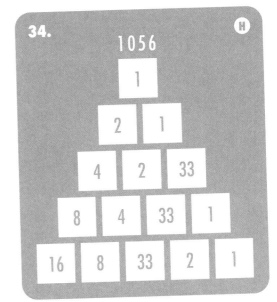

35.

39^5

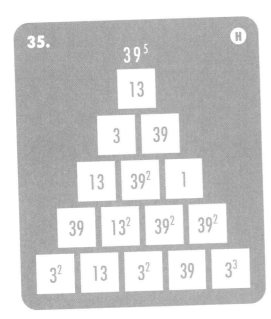

36.

1001

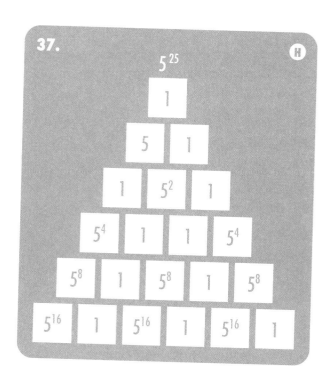

37.

5^{25}

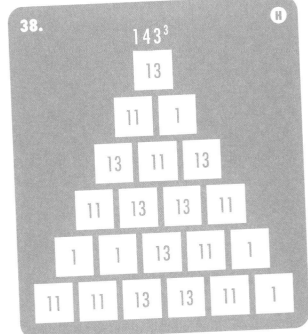

38.

143^3

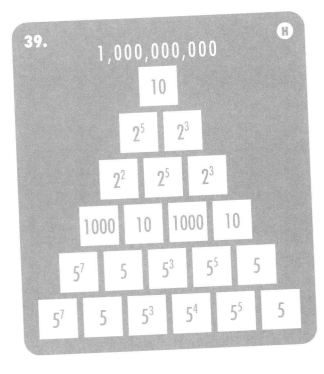

39.

1,000,000,000

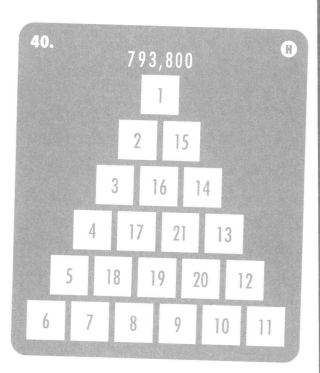

40.

793,800

PYRAMID DESCENT
STRATEGIES

1. **Find the prime factorization of the target product.**

   It is easier to keep track of factors than it is to find the product of every possible path.

   We can use a factor tree to find the prime factorization of the target product.

   The prime factorization of 60 is $2^2 \times 3 \times 5$. So, we look for a path whose factors include two 2's, one 3, and one 5.

   For the more difficult prime factorizations, it may help to look at the numbers in the pyramid for clues about what factors might be used.

2. **Use divisibility tests when finding the prime factorization of the target product.**

   Use the following divisibility tests to find prime factorizations of numbers. A number is divisible by:

   2: If its ones digit is even.

   3: If the sum of its digits is divisible by 3.

   4: If the number formed by its last two digits is divisible by 4.

   5: If its ones digit is 0 or 5.

   6: If it is divisible by both 2 and 3.

   9: If the sum of its digits is divisible by 9.

   10: If its ones digit is 0.

   (Find out why all of these tests work and how to use them in Beast Academy 4B and 4C.)

3. **Cross out numbers that cannot be on the path.**

   If a number is not a factor of the target product, cross it out.

   Which numbers in the pyramid below cannot be part of the path?

   The prime factorization of 270 is $2 \times 3^3 \times 5$. So, the path must include exactly one 2.

   Since the path begins with a 2, it cannot contain any more 2's. So, we can cross out every number that is a multiple of 2.

4. **Circle numbers that must be on the path.**

   Some factors needed for the target product only appear in a few blocks.

How can we include the two 7's needed in the path for the puzzle below?

Sevens only appear in two rows. So, the path must include one 7 in the second row and one 7 from the bottom row.

There is only one 7 in the second row, so we circle it.

Then, we circle the only 7 that we can reach with a path from the 7 in the second row.

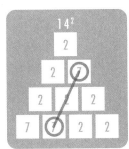

5. Work backwards.

Sometimes it is easier to start at the bottom of the pyramid and find a path to the top.

How does the path end in the puzzle below?

The prime factorization of 1,440 is $2^5 \times 3^2 \times 5$. In the bottom row, 11, 13, and 14 are not factors of 1,440, so we can cross these out.

If the path ends at 15, it must also include the 10 above it. Since 1,440 only has one 5 in its prime factorization, it cannot have both a 15 and a 10 on the path. So, the path must end at 12.

Then, which of the two squares above the 12 must be part of the path?

DUTCH LOOP

DIFFICULTY LEVEL:

★ — ★ ★ ★ ★ ★

In a **Dutch Loop** puzzle, the goal is to draw a single looped path that passes through every square exactly once. The path can only travel horizontally and vertically, and cannot branch.

The path must:
Travel straight through any square with a white circle (○).
Turn at any square with a green circle (●).

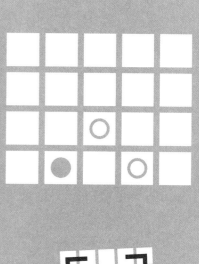

There is only one way the path can pass through each corner. We draw the path through the corners as shown.

Then, in the bottom-left corner, the path must turn up at the green circle.
In the bottom-right, the path must continue straight through the white circle.

From the highlighted square, the path must turn up and pass through the white circle above.

Now, the two ends of the path in the lower-left corner cannot connect to each other, so they must continue up.

The two ends in the top-left corner cannot connect, so both must continue to the right as shown.

Finally, the path must visit the two remaining squares. We complete the loop.

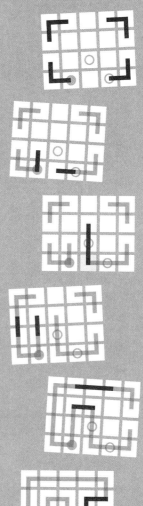

9.

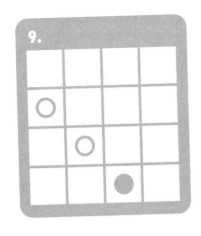

10.

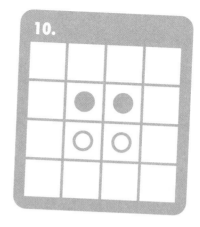

11.

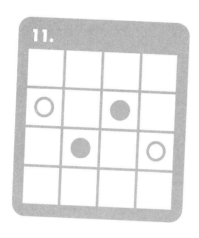

12.

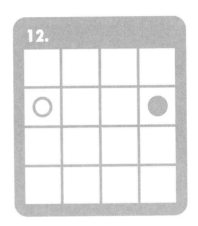

13.

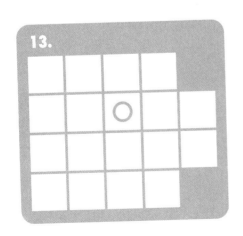

14.

15.

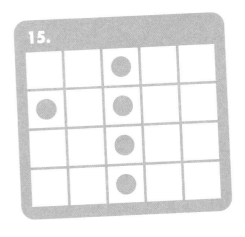

16.

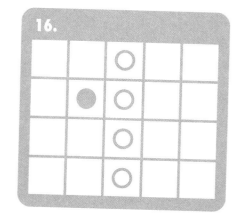

17.

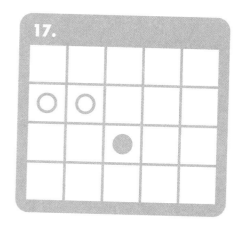

18.

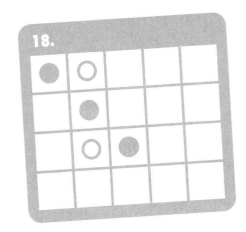

19.

20.

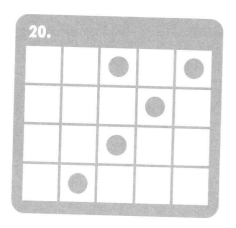

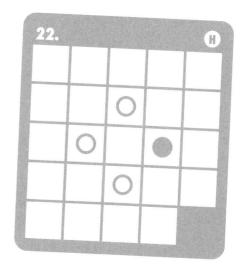

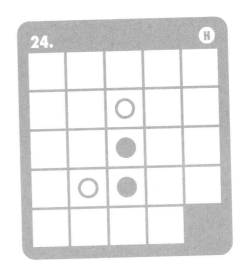

ℍ Hints on pages 173-174

Beast Academy Puzzles 4

27.

28.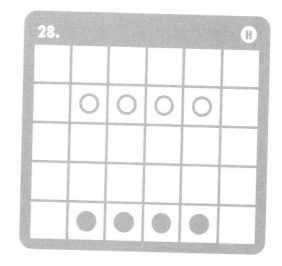

29.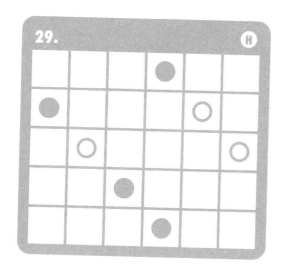

30.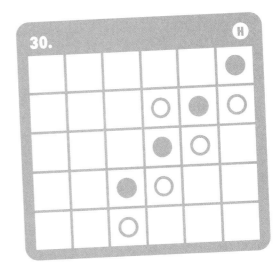

31.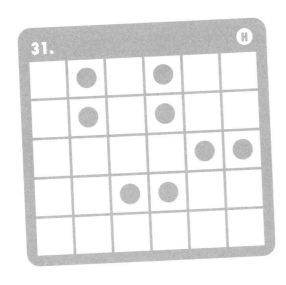

32.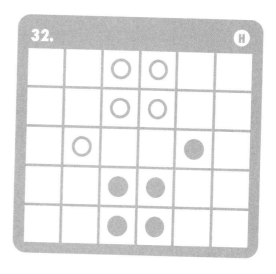

H Hints on pages 173-174

33.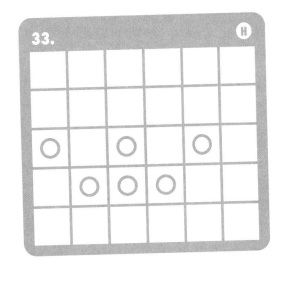

34.

35.

36.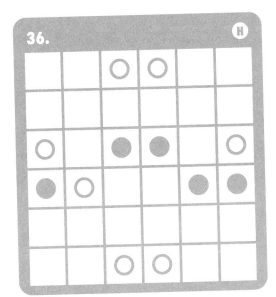

37.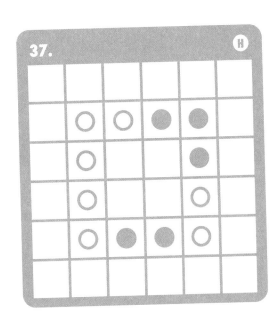

38.

H Hints on pages 173-174

Beast Academy Puzzles 4

39.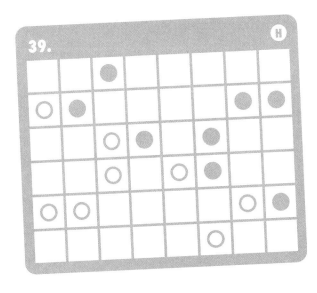

40.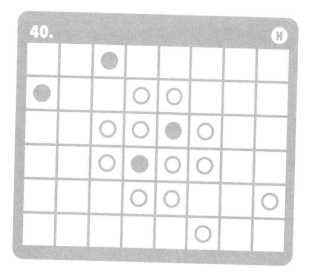

41.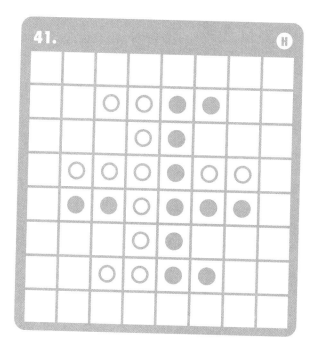

42.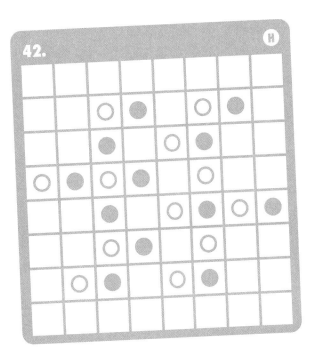

H Hints on pages 173-174

DUTCH LOOP
STRATEGIES

1. Find known parts of the path.

Instead of trying to draw the path as one continuous loop from the start, we can find the pieces of the path we know and then work to connect them.

2. Look at the corners.

There is only one way the path can pass through a corner.

How can we draw the path in the corners of the puzzle below?

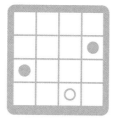

There is only one way for a path to enter and exit each corner. We draw the parts of the path that pass through the corners as shown.

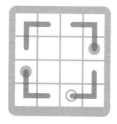

3. Draw a path through any ◯ with only one choice.

If there is only one way to draw a path through a ◯, draw it.

How can the path pass through each ◯ below?

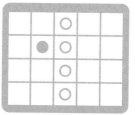

In the top and bottom rows, there is only one way to draw a path through each ◯.

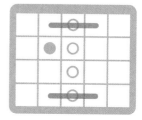

Then, there is only one way to draw the path through the other two ◯s.

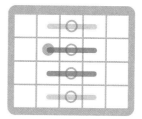

4. Look for "loose ends" that can only be extended one way.

Sometimes there is only one way to extend a path.

How can we extend each end of the partial path in the bottom-left corner?

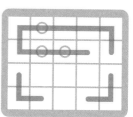

There is only one way to extend the path from the highlighted square below.

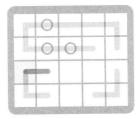

Then, both ends of the path can only continue to the right.

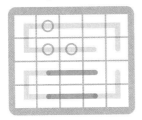

5. Don't create partial loops.

How can the path be drawn through the ○s in the puzzle below?

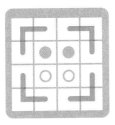

If we draw the path horizontally through the ○s, it splits the path at the bottom half of the grid into its own loop.

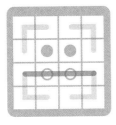

So, we must draw paths vertically through both ○s.

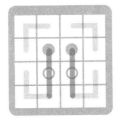

6. Don't isolate squares.

The path must enter and exit each square. So, each square must have at least two possible connections.

How can we extend the path from the highlighted square?

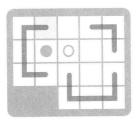

If we extend the path down, it must turn at the ● and continue straight through the ○. But, this makes it impossible to connect the highlighted square to the path.

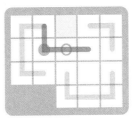

So, we must extend the path to the right.

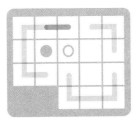

7. Draw part of the path through a ⬤.

The path through each ⬤ can extend either up or down, but not both.

Similarly, the path through each ⬤ can extend either left or right, but not both.

What part of the path through the highlighted ⬤ below can we draw?

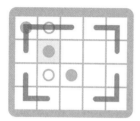

The path through the highlighted ⬤ cannot extend up, so it must extend down and through the ◯ below it.

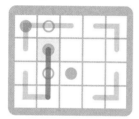

8. Draw path pieces through squares that have only two open sides.

If a square has only two open sides, it must enter and exit through those sides.

How will the path travel through the highlighted square in the puzzle below?

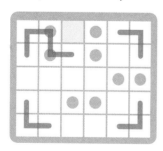

The highlighted square has only two open sides. So, the path must pass through it as shown.

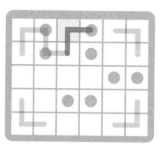

9. Draw walls.

If a path cannot connect two squares that touch, we can draw a wall between them.

What walls can be drawn in the puzzle below?

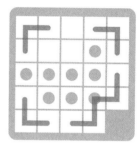

The path passes through two sides of the square below. So, we can draw walls around the other two sides.

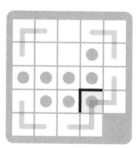

The path from the highlighted ● below cannot extend left, so it must extend right (strategy 7).

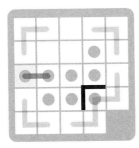

This path cannot continue to the right, so we can draw a wall to its right.

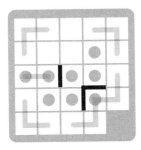

Similarly, the path from the highlighted ● below cannot extend left, so it must extend right, and we can place a wall to its right.

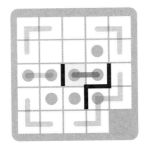

When walls create new corners, we can apply strategy 2.

(Now, which parts of the path can we complete?)

HIVE

DIFFICULTY LEVEL:

★—★★★★★

In a **Hive** puzzle, fill every hexagon with a number from 1 to 7 so that every number is the smallest number it doesn't touch.

In other words, each number:

- Must touch every number less than it.
- Cannot touch a number equal to it.
- May touch numbers greater than it.

For example, every 4:

- Must touch a 1, 2, and 3.
- Cannot touch a 4.
- May touch 5's, 6's, and 7's.

The bottom-right hexagon touches a 1 and a 2, but not a 3. So, we place a 3 in the bottom-right hexagon.

The 2 on the right must touch a 1. So, we place a 1 in the only empty hexagon it touches.

The top-left 2 must touch a 1. There is only one place to put a 1 that does not touch another 1.

The empty hexagon in the top row touches a 1, a 2, a 3, and a 4, but not a 5. So, we place a 5 in it.

The empty hexagon in the bottom row touches a 1, but not a 2. So, we place a 2 in it.

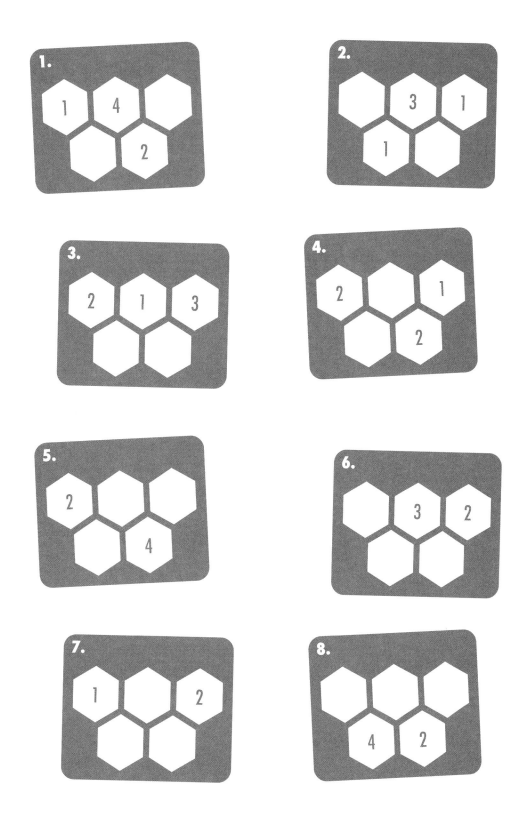

9.

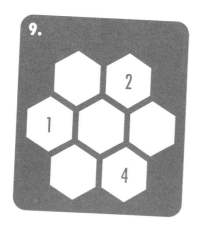

10.

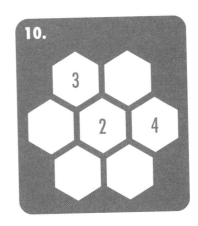

11.

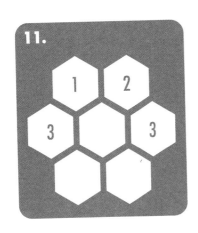

12.

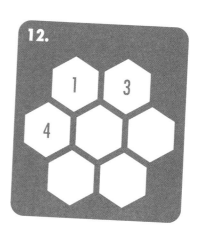

13.

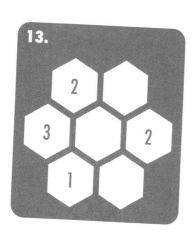

14.

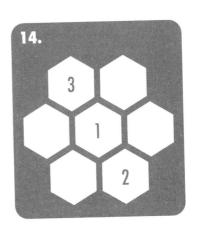

15.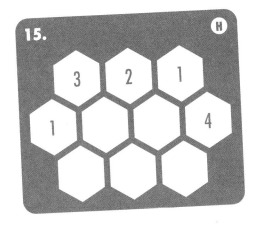

16.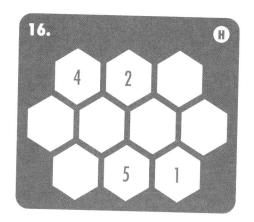

17.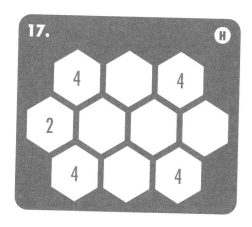

18.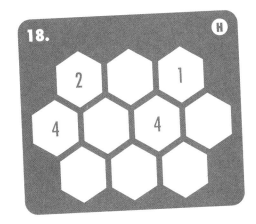

19.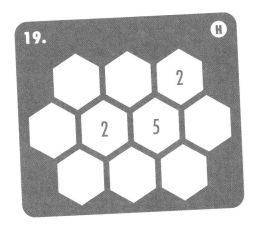

20.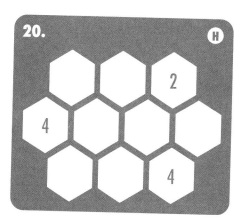

H Hints on pages 175-177

21.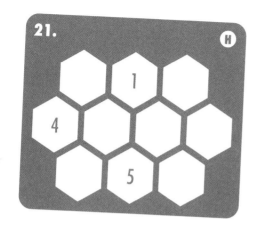

22.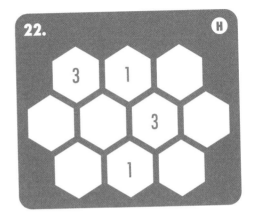

23.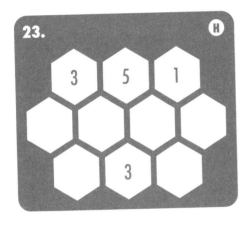

24.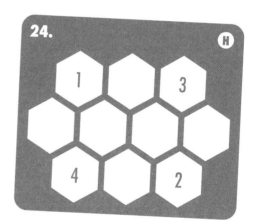

25.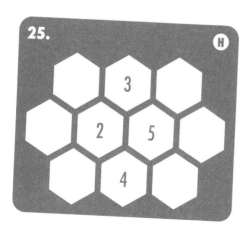

26.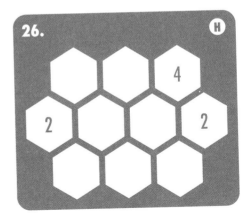

27.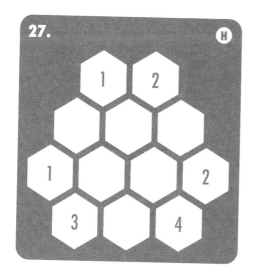

28.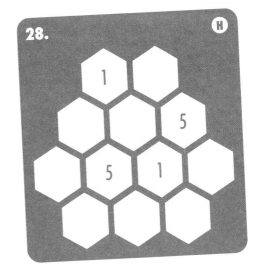

29.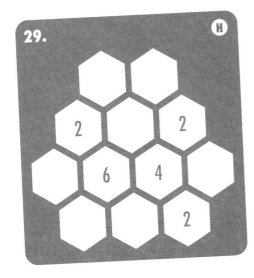

30.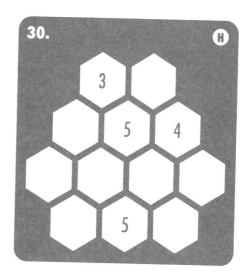

31.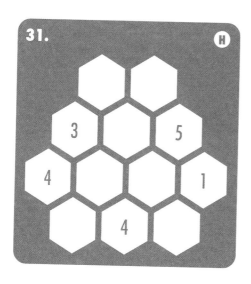

32.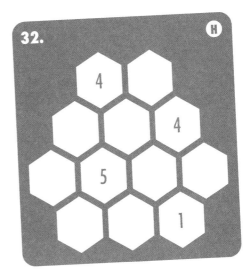

H Hints on pages 175-177

33.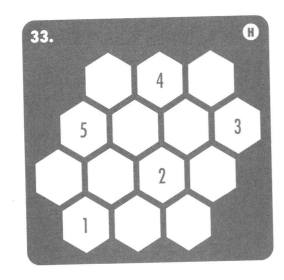

34.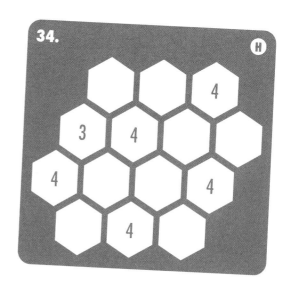

35.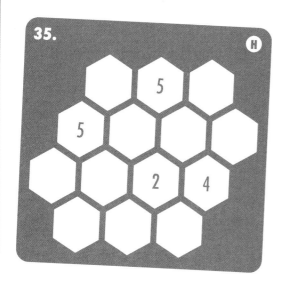

36.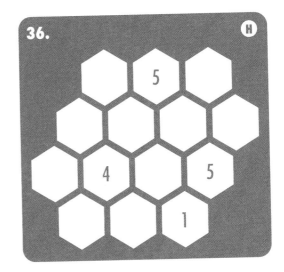

37.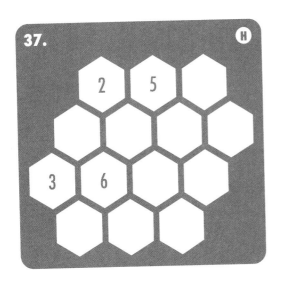

38.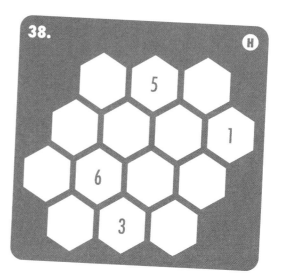

39.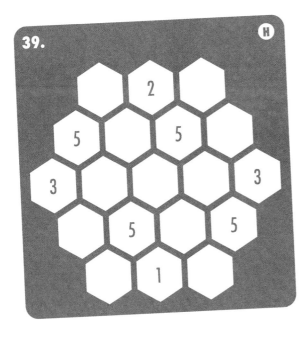

40.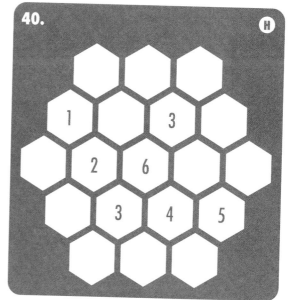

41.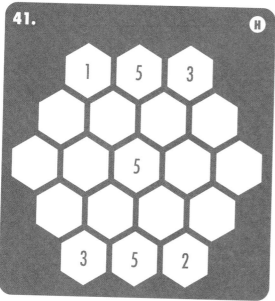

42.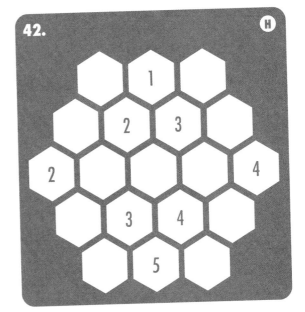

43.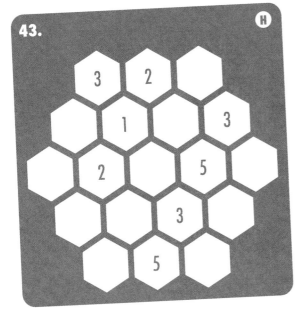

44.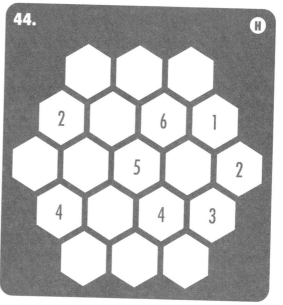

Hints on pages 175-177

45.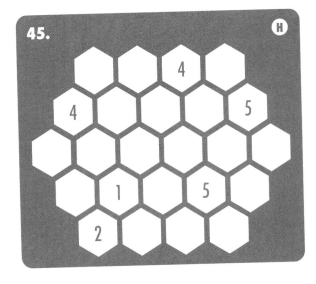

46.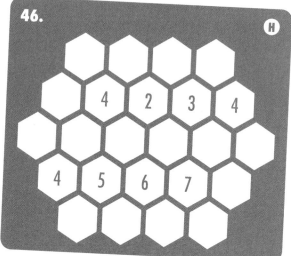

47.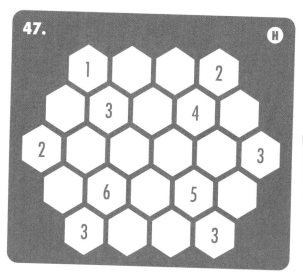

48.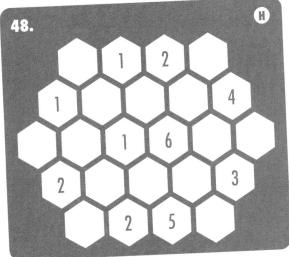

H Hints on pages 175-177

Beast Academy Puzzles 4

49. H

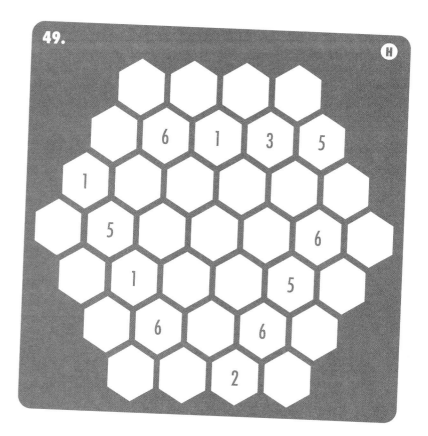

50. H

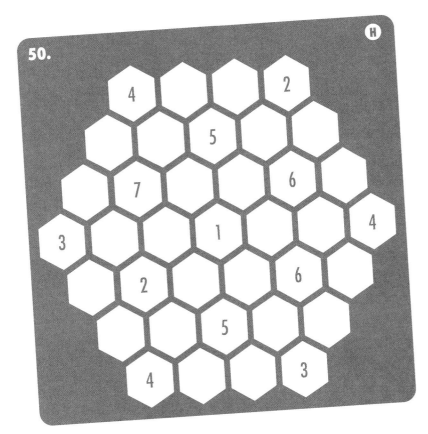

H Hints on pages 175-177

HIVE
STRATEGIES

1. Fill empty hexagons that are completely surrounded by numbers.

If an empty hexagon is completely surrounded by numbers, we can fill it with the smallest number it doesn't touch.

What numbers fill the empty hexagons below?

The empty hexagon in the bottom row touches a 1, a 2, and a 4. The smallest number it does not touch is a 3.

The empty hexagon on the right touches a 2 and a 4. The smallest number it does not touch is a 1.

2. Find the numbers that are still needed around a hexagon.

Knowing which numbers need to touch a hexagon helps us fill the empty hexagons that surround a number.

Which numbers must touch the 4?

The 4 must touch a 1, a 2, and a 3. Since there are only three empty hexagons that touch the 4, these hexagons must be filled with a 1, a 2, and a 3.

We use strategy 3 below to fill these hexagons.

3. Two of the same number can't touch.

Where can we place the 1, 2, and 3 that touch the 4 below?

The 1 cannot touch a 1, and the 2 cannot touch a 2. There is only one place to put the 1 where it does not touch a 1, and one place to put the 2 where it does not touch a 2.

Beast Academy Puzzles 4

Now, there is only one place to put the 3.

4. Place 1's.

Every hexagon that is not a 1 must touch a 1. So, every hexagon must either touch a 1 or be a 1.

Where can we place 1's in the puzzle below?

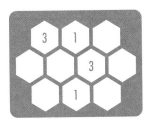

We can't place a 1 in any of the highlighted hexagons, since all of these hexagons already touch a 1.

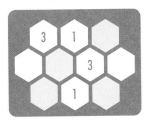

Since the leftmost hexagon cannot touch a 1, it must be a 1. Similarly, the rightmost hexagon cannot touch a 1, so it must be a 1.

5. Look at pairs of empty hexagons together.

Knowing what numbers can fill one hexagon can help us fill its neighbor.

What numbers fill the two hexagons below?

Since the lower hexagon already touches a 1 and a 2, it must be at least a 3.

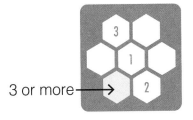

3 or more

So, the hexagon above it touches a 1, but not a 2. So, we place a 2 in the leftmost hexagon.

Then, the smallest number the lower-left hexagon does not touch is a 3.

6. Place large numbers.

Large numbers must be surrounded by many other numbers, so options are limited for placing large numbers.

Where can we place the 4 that touches the 5?

The 4 must touch a 1, a 2, and a 3. If we place a 4 in any of the hexagons highlighted below, we won't be able to surround it with a 1, a 2, and a 3.

There is only one hexagon that touches the 5 that can be surrounded with a 1, a 2, and a 3. We place the 4 in this hexagon.

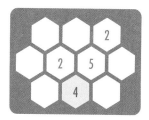

7. Combine clues.

Multiple clues can apply to the same hexagon.

What number goes in the highlighted hexagon below?

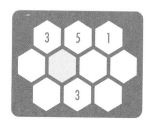

The 3 in the top row must touch a 1 and a 2. The 5 in the top row must touch a 2 and a 4.

So, the only number we can place in the hexagon that they both touch is a 2.

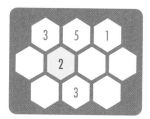

8. Guess and check.

If there are limited options for a hexagon, try them all. Eliminate the choices that don't work.

How can we fill the hexagons that touch the 4?

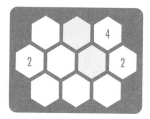

The empty hexagons must contain a 1 and a 3. We first try placing the 3 on top and the 1 below.

Now, the 3 must touch a 2. There is no way to do this without having the 2 touch another 2.

So, we must place the 1 and the 3 as shown below.

9. Keep looking!

If we're stuck in one place, we can try looking at another part of the puzzle. Keep looking for clues to find the easiest next step.

FACTOR CAVE

DIFFICULTY LEVEL:

In a **Factor Cave** puzzle, the goal is to shade some of the squares of a grid to map the walls of a cave. Squares with numbers cannot be shaded.

Each number clue gives a product: the number of squares it can "see" in its row times the number of squares it can "see" in its column.

A number can see all unshaded squares (including itself) that are not blocked by shaded squares vertically and horizontally.

First, squares with numbers cannot be shaded, so we circle them to show they must be unshaded.

Next, we look at the 16 clue. There are several ways to write 16 as the product of two numbers: 16×1 (or 1×16), 8×2 (or 2×8), and 4×4. The only option that fits in this grid is 4×4. So, the 16 can see 4 squares in its row and 4 squares in its column.

We circle these squares to show they must be unshaded.

Next, 9 can be written as 9×1 (or 1×9) or 3×3, but only 3×3 fits.

The only way that the 9 can see 3 squares in its row and 3 squares in its column is shown on the right.

We finish the puzzle as shown and check our work:

- The 6 can see 3 squares in its row and 2 squares in its column. 3×2=6 ✓

- The 2 can see 1 square in its row and 2 squares in its column. 1×2=2 ✓

7.

8.

9.

10.

11.

12.

13. H

1		2	
	1		2
1		2	
			2

14. H

3			3
	4	4	

15. H

	2		2
		2	
2			
2			2

16. H

2		3	
			3
	2		

17. H

3		4	
	3		4

18. H

12	6		
		8	8

H Hints on pages 178-179

19.

20.

21.

22.

23.

24.

25.

26.

27.

28.

29.

30.

H Hints on pages 178-179

31.

32.

33.

34.

35.

36.

41.

42.

43.

44.

45.

46.

Ⓗ Hints on pages 178-179

FACTOR CAVE
STRATEGIES

1. Circle squares that must be unshaded.

This makes it easier to keep track of our progress on a puzzle.

Don't forget that all clue numbers must be unshaded, so we can always circle them.

2. Look for clues that are "done".

In the puzzle below, which clues already equal the product of the number of squares they can see in their row and column?

The 2's highlighted below can already see 2 squares in their column and 1 square in their row. Since 2×1=2, these clues are "done".

These clues cannot see any more squares, or else their product would be greater than 2. So, the squares surrounding them must be shaded.

3. Look at numbers with few possible factor pairs.

Some clue numbers only have one possible factor pair.

How many squares can the 9 see in its row and column in the puzzle below?

9 can be written as the product of two numbers in three ways: 9×1, 3×3, and 1×9. However, the grid is only 4 squares tall and wide, so it is impossible to see 9 squares in a row or column.

Therefore, the 9 must see 3 squares in its row and 3 squares in its column.

4. Look at numbers on the edges or corners.

From strategy 3, we know that the 9 in the puzzle below sees 3 squares in its row. Which squares in the row must be shaded and unshaded?

Because the 9 is on the left edge, it can only see squares to its right. In order to see 3 squares in its row, the 2 squares to the right of the 9 must be unshaded, and the last square must be shaded.

5. Find squares that must be unshaded, even if there isn't a unique way to factor a number.

Which squares in the same row and column as the 12 must be unshaded?

12 sees either 3 squares in its row and 4 squares in its column, or 4 squares in its row and 3 squares in its column.

Either way, the 12 will always see *at least* 3 squares in its row and column, so all of the squares below must be unshaded.

This strategy works best with numbers on the edges or corners, but we can also use it with numbers that are close to the edge.

6. Look at large numbers.

Larger numbers are generally more helpful because they require more unshaded squares.

In the puzzle below, which squares must be unshaded for the 12 clue?

The valid factor pairs for the 12 are 3×4 and 4×3. So, the 12 must see at least 3 squares in its row. If we shade the highlighted square below, then the 12 sees at most 2 squares in its row.

So, the square to the left of the 12 must be unshaded.

We can use the same logic with the 12's column to see that the square below the 12 must also be unshaded.

7. Check which possible factor pairs and orientations fit.

If a number has several possible factor pairs, try pairs and eliminate the ones that don't work.

It is often helpful to start with numbers in the corners (strategy 4).

In this puzzle below, there are two options. Either the 3 in the top-left corner sees 3 squares in its row and 1 square in its column, or it sees 1 square in its row and 3 squares in its column.

Which option works?

If the 3 clue sees 3 squares in its row, then the 2 squares to the right of the clue must be unshaded:

However, the top-right corner must also be unshaded because it has a clue, so the 3 clue now sees four squares in its row. This is not possible!

So, the 3 clue sees 1 square in its row and 3 squares in its column.

This strategy can also be useful for numbers on the edges (Strategy 4) or for numbers close to other numbers (Strategy 8).

8. If two clues can see each other, then they must share a common factor.

How many squares can the highlighted clues see in their row?

Because the 4 and 6 see each other horizontally, they must see the exact same number of squares in their row. This number must be a factor of both 4 and 6, so it is either 1 or 2.

Since these clues already see two squares in their row, the 4 and 6 must see exactly 2 squares in their row.

That means the squares below must be shaded.

Also notice that if two clues do not share any common factors, then they cannot see each other.

This strategy can be helpful whenever there are two clues close to each other.

FACTOR BLOBS
DIFFICULTY LEVEL:
★—★★★★

In a **Factor Blob** puzzle, the goal is to circle "blobs" of factors whose product is a given target number.

Every square in a blob must share at least one edge with another square in the blob. Blobs may not overlap. The goal is to use every number in the grid in a blob.

Product: 30

6	3	5
5	2	3
2	15	10

Since 6×5=30, the 6 in the top left corner must be paired with a 5. So, we pair the 6 with the 5 below it in a blob.

Similarly, since 10×3=30, the 10 in the bottom right corner must be paired with the 3 above it.

Product: 30

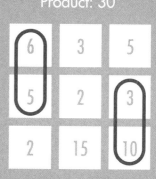

We then group the remaining numbers into blobs as shown so that the product of the numbers in each blob is 30.

Product: 30

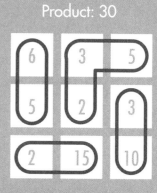

1. Product: 18

9	6	3
2	2	9
3	2	3

2. Product: 27

3	3	3
3	9	3
3	9	9

3. Product: 33

11	1	11
11	3	1
1	3	3

4. Product: 30

6	5	3
2	15	10
5	2	3

5. Product: 63

7	9	3
21	7	1
3	3	1

6. Product: 20

2	5	5
2	2	2
5	1	4

7. Product: 12

3	4	3
2	2	4
6	3	2

8. Product: 21

7	1	7
7	1	3
3	1	3

9. Product: 80

10	2	2
2	20	1
2	4	40

10. Product: 15

3	1	3
5	5	5
5	3	3

11. Product: 50

5	2	2
2	25	5
5	5	1

12. Product: 8

4	4	2
2	2	2
2	2	4

13. Product: 56

7	14	2
2	2	2
2	28	2

14. Product: 16

8	1	2
4	2	4
2	4	1

15. Product: 24

1	2	3
4	6	8
3	2	2

16. Product: 32

4	2	8
2	2	16
1	4	2

17. Product: 40

5	2	2
1	10	4
2	5	8

18. Product: 36

6	3	6
1	2	3
6	2	6

19. Product: 75 H

5	1	1	3
5	1	1	3
5	5	25	3

20. Product: 28 H

2	2	2	2
7	7	7	7
4	7	4	4

21. Product: 84 H

2	7	2	21
1	2	2	2
2	3	3	7

22. Product: 81 H

9	3	3	3
3	9	1	9
9	3	9	3

23. Product: 60 H

5	1	1	1
5	2	4	1
3	1	1	6

24. Product: 36 H

12	3	1	2
3	3	1	3
2	3	1	4

H Hints on page 180

☆ ☆ ★ ★ ★

25. Product: 78 **H**

2	2	1	13
1	39	3	13
26	3	2	1
13	2	3	3

26. Product: 60 **H**

1	2	3	4
2	3	4	5
3	4	5	6
5	3	5	10

27. Product: 22 **H**

2	2	1	11
1	2	1	1
1	2	2	11
11	11	1	11

28. Product: 210 **H**

7	10	1	1
1	3	5	7
1	3	3	7
10	2	1	1

29. Product: 140 **H**

7	2	1	5
2	7	5	1
2	5	7	4
5	4	2	7

30. Product: 165 **H**

11	5	3	11
3	5	3	5
11	3	3	5
11	1	5	11

31. Product: 135

1	3	3	9
1	5	5	3
1	5	5	9
3	3	9	3

32. Product: 96

12	2	2	4
12	2	2	2
2	2	2	24
2	2	3	2

33. Product: 216

6	1	9	6
4	6	3	1
2	6	6	1
6	1	6	36

34. Product: 384

1	2	1	4
2	2	4	4
6	6	8	8
6	1	8	1

35. Product: 100

2	50	2	4
2	25	5	25
2	10	2	5
5	10	4	5

36. Product: 1,000

2	50	2	4
2	25	5	25
2	10	2	5
5	10	4	5

Hints on page 180

37. Product: 360 **H**

2	5	5	4	18
2	9	2	4	18
10	9	2	3	20
10	6	6	3	20

38. Product: 330 **H**

11	2	3	5	11
5	3	5	11	3
2	11	3	2	5
5	2	11	2	3

39. Product: 81 **H**

27	3	3	3	9
3	27	3	27	3
3	3	27	3	9
3	27	3	3	3

40. Product: 150 **H**

5	5	5	5	5
6	1	3	1	5
1	1	6	1	5
10	3	1	1	2

☆ ★ ★ ★ ★

41. Product: 340 H

17	2	2	4	17
20	5	5	5	5
17	5	17	5	17
2	1	4	1	4
17	2	2	2	17

42. Product: 48 H

8	1	1	1	2
1	1	16	1	1
1	1	1	12	1
1	6	1	1	3
24	1	1	1	4

43. Product: 128 H

2	2	2	2	2
2	64	64	16	2
2	32	2	4	2
2	32	16	1	2
2	2	2	2	2

44. Product: 240 H

2	2	3	4	5
2	3	4	5	4
3	4	60	1	2
4	5	4	1	6
5	4	10	6	2

H Hints on page 180

FACTOR BLOBS
STRATEGIES

1. Factor the products.

Before starting, write down the prime factorization of the target product and the numbers in the grid. See Pyramid Descent strategies 1 and 2 on page 54 for details.

2. Start at the corners.

Numbers in the corners have limited options. As we create blobs, we create more "corners".

Which corner number can only be part of one blob?

The 8 in the top-right corner only needs two more factors of 2 to give a product of 32. There is only one way we can draw a blob with the 8 that gives a product of 32.

3. Draw walls.

If two touching squares can't be part of the same blob, draw a wall between them.

Which walls can we draw between numbers in the puzzle below?

The target product is 50=2×5×5.

Since the target product only has one 2, we draw walls between the pair of 2's.

Since the target product only has two 5's, we draw walls between 5 and 25, since no blob can have three 5's.

(Then, what blob includes the top-right 2?)

4. Start with numbers in the grid that have lots of factors.

If a number in the grid has most of the target product's factors, then the numbers that can be in its blob are limited.

Which number in the grid has the most factors?

The target product 40=2×2×2×5 has four prime factors.

In the grid, 10 and 4 each have two prime factors.

8=2×2×2 has three prime factors, so it only needs a 5 to be a completed blob.

5. Don't isolate regions that can't form blobs.

When drawing blobs, don't separate numbers or groups of numbers that can't form blobs.

Which 3 is part of the same blob as the top-left 7?

Since 21=3×7, each blob must have one 3 and one 7.

If we connect the 7 to either of the 3's below, then one of the other 7's can't be paired with a 3.

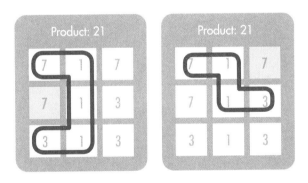

So, the top-left 7 must be part of the same blob as the bottom-right 3.

6. Find pairs of factors that must be linked.

Sometimes pairs of numbers in the grid must be in the same blob.

In the puzzle below, what blob is the 9 part of?

The 9 needs to be connected to one 3 and three 2's. The only numbers that give us 2's without more 3's are 2 and 4. Since we need at least three more 2's (and we can't connect to more than 1 more 3), the 9 *must* be connected to the 4.

There is only one way to connect the 9 and 4 that gives a product of 216 and avoids isolating squares.

7. Find factors that only appear once in the target's prime factorization.

If a prime factor only appears once in the target product, then all of the squares in the grid that include that factor will be in different blobs.

In the puzzle below, the target product 96=2×2×2×2×2×3 has one factor of 3. Which squares have a factor of 3?

There are a total of four numbers in the grid with a factor of 3: 12, 12, 24, and 3.

We highlight these factors to see that there are a total of four blobs in the puzzle.

Then, which of these blobs can contain the 4 in the top-right corner?

8. Guess-and-check.

If there are a small number of ways to make a blob, test them all.

Which blob includes the 39?

Product: 78			
2	2	1	13
1	39	3	13
26	3	2	1
13	2	3	3

The target product is 78=2×3×13. So, the blob with 39=3×13 only needs one factor of 2.

We can connect the 39 to either 2 in the top row. We try connecting the 39 to the 2 above it.

Product: 78			
2	2	1	13
1	39	3	13
26	3	2	1
13	2	3	3

Then, the top-left 2 must be part of the same blob as 26=2×13. However, this creates a blob with two factors of 2, which is too many.

So, 39 can only be in the same blob as the top-left 2.

FRACTION SUMDOKU

DIFFICULTY LEVEL:
★★ – ★★★★★

In a **Fraction Sumdoku** puzzle, the goal is to fill the grid with the given digits according to the following rules:

- Each digit must appear exactly once in every row and every column.

- Squares with a slash (/) must contain two digits that form a fraction (for example, 3/4 or 2/1).

- The numbers in each outlined cage must add up to the sum given in its top-left corner. Cages with no number can have any sum.

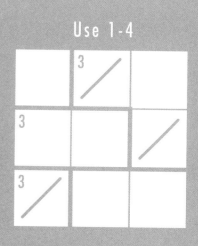

Use 1-4

In the lower-left cage, the only fraction we can write using the digits 1 through 4 that equals 3 is 3/1.

In the cage in the middle row, the only way to get a sum of 3 is using 1 and 2. Since the left column already has a 1, we write a 2 on the left and a 1 on the right.

We complete the left column with a 4 in the top-left corner.

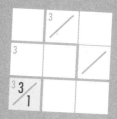

The remaining digits in the middle row are 3 and 4. So, the fraction in the middle row can be 3/4 or 4/3.

If we use 3/4 in the right column, the other fraction in its cage must be 1/4 to make the sum a whole number.

Since there is already a 4 in the top row, this does not work.

So, we use 4/3 in the middle row and 2/3 for the other fraction in its cage.

From there, we complete the puzzle as shown.

1.
Use 1-3

3

3

2.
Use 1-3

8   2

3.
Use 1-3

3   4

Some cages don't have a number in the top-left corner.

These cages can have any sum, so you'll need to use the other clues to figure out what goes in them!

4.
Use 1-4

4   16   2

2   3

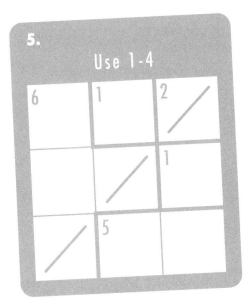

5.
Use 1-4

6   1   2

1

5

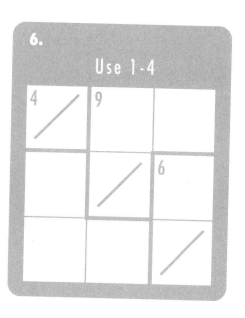

6.
Use 1-4

4   9

6

7. H

Use 1-4

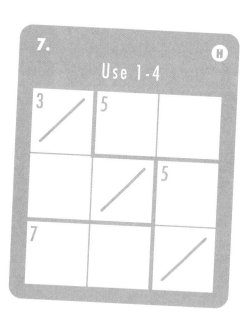

8. H

Use 1-4

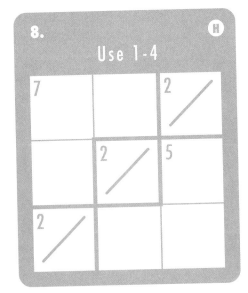

9. H

Use 1-4

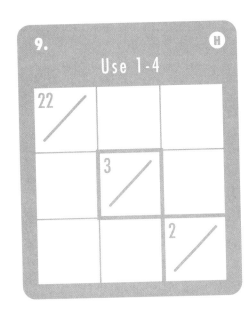

10. H

Use 1-4

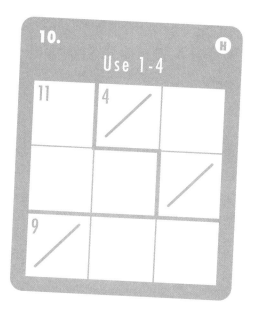

11. H

Use 1-4

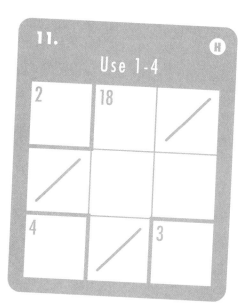

12. H

Use 1-4

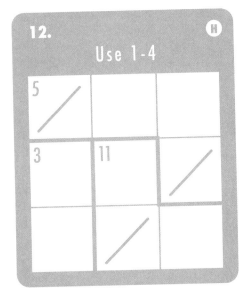

13.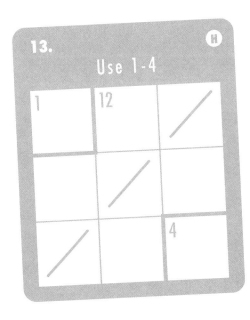

Use 1-4

14.

Use 1-4

15.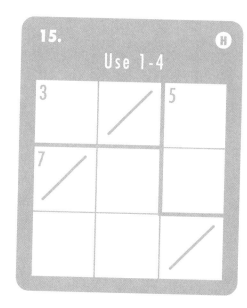

Use 1-4

16.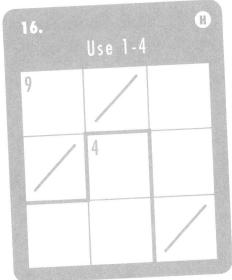

Use 1-4

17.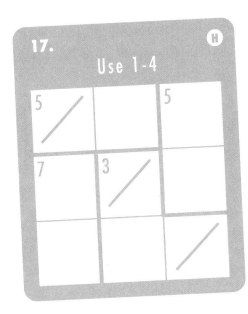

Use 1-4

18.

Use 1-4

19.

Use 1-5

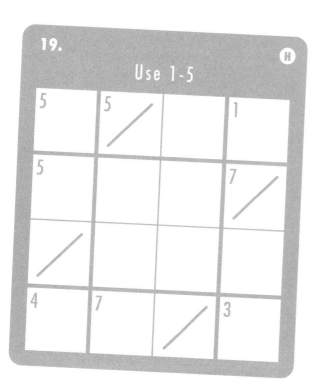

20.

Use 1-5

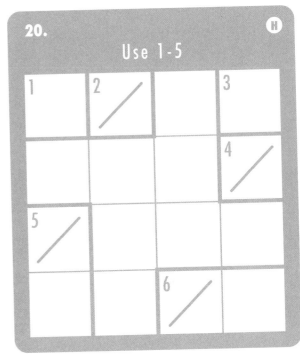

21.

Use 1-5

22.

Use 1-5

☆ ★ ★ ★ ★

23. Ⓗ

Use 1-5

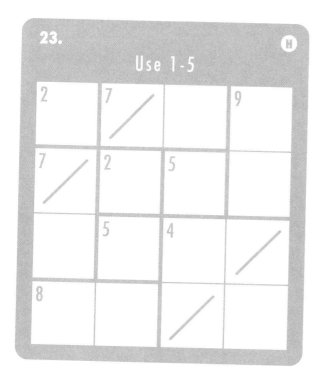

24. Ⓗ

Use 1-5

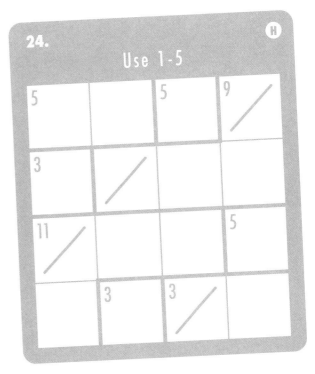

25. Ⓗ

Use 1-5

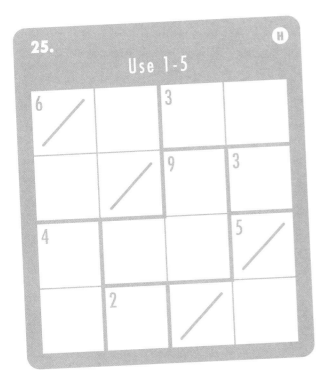

26. Ⓗ

Use 1-5

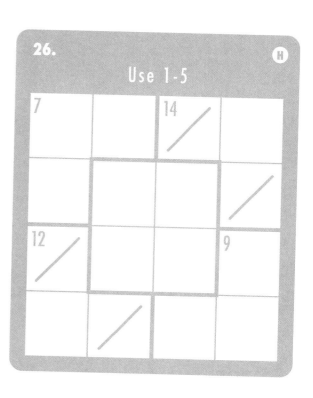

Ⓗ Hints on pages 181-182

27. Ⓗ

Use 1-5

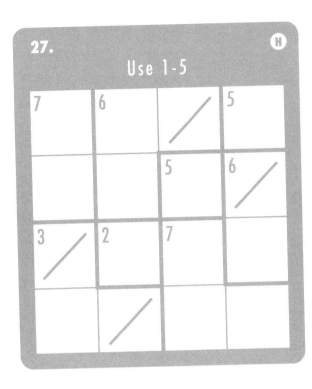

28. Ⓗ

Use 1-5

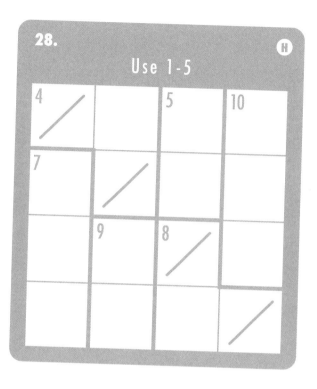

29. Ⓗ

Use 1-5

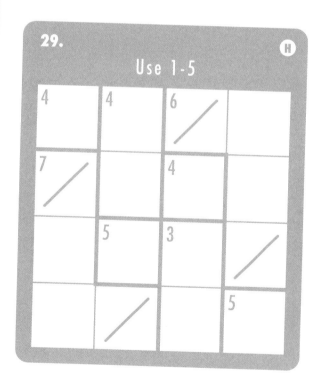

30. Ⓗ

Use 1-5

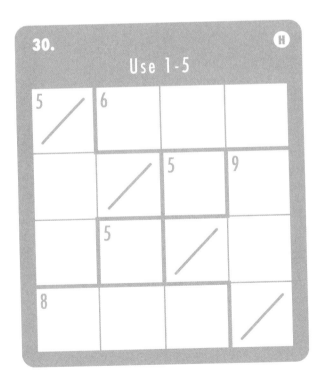

31.

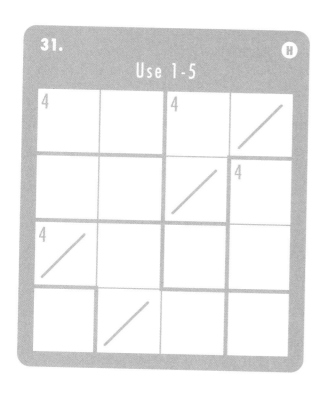

32.

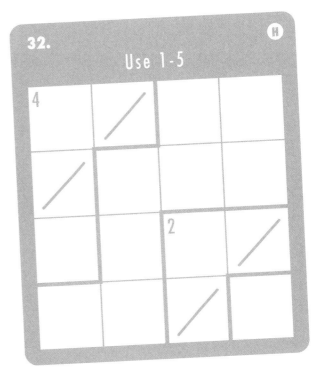

33.

34.

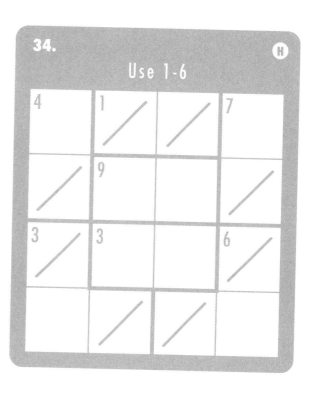

H Hints on pages 181-182

FRACTION SUMDOKU

★ ★ ★ ★ ★

35.

Use 1-6

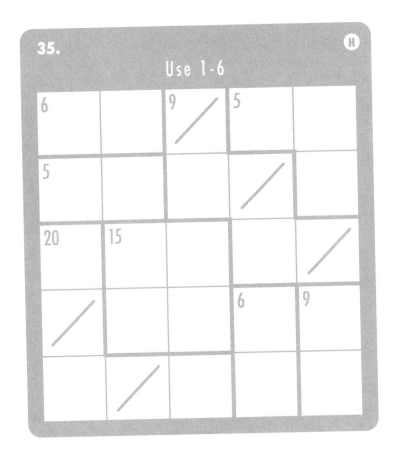

36.

Use 1-6

Hints on pages 181-182

Beast Academy Puzzles 4

37.

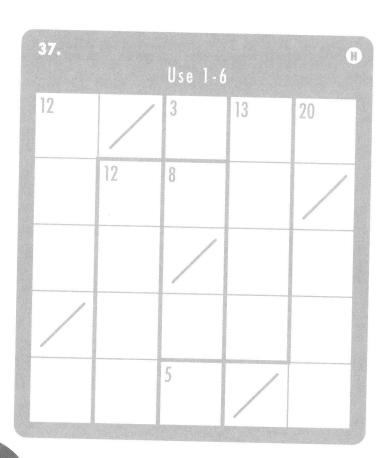

Use 1-6

12		3	13	20
	12	8		
		5		

This one stumped Captain Kraken!

38.

Use 1-7

1	4	8		
		14	12	
17				
			4	
7				4

H Hints on pages 181-182

FRACTION SUMDOKU

STRATEGIES

1. Fill cages that only have one square.

If a cage has only one square, we can often complete it.

What numbers fill the corners?

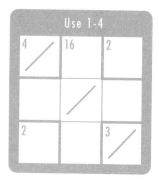

The bottom-left and top-right cages must each have a 2.

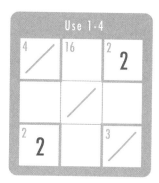

In the top-left cage, the only fraction that equals 4 and uses the digits 1-4 is 4/1.

Similarly, we can only place 3/1 in the bottom-right corner.

2. Complete rows, columns, and cages that are only missing one digit.

How can we complete the 5 cage?

The right column already uses a 1, a 2, and a 4, so we place a 3 in the bottom-right corner.

Then, we complete the 5 cage by placing a 2 as shown.

3. Look for digits that have limited placement options.

Where can we place the other 4's?

The top row needs a 4. There is already a 4 in the right column, so we can't place a 4 in the top-right square. So, we must place a 4 in the top-center square.

Similarly, in the left column, we can only place a 4 in the center-left square.

4. Find cages with limited possibilities.

Sometimes it helps to know which digits are needed for a cage even if we don't know which square each digit fills.

What digits fill the 7 cage?

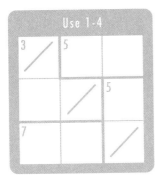

The only pair of digits between 1 and 4 that has a sum of 7 are 3 and 4. So, we must fill the 7 cage with 3 and 4 in some order.

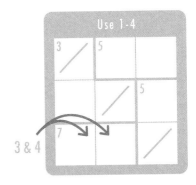

5. If a cage has one fraction, that fraction must equal a whole number.

Since the sum of a cage is a whole number, if there is only one fraction in a cage, it must equal a whole number.

What fraction fills the bottom-right corner?

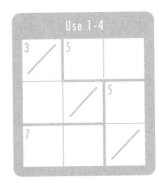

In the 7 cage, we can only get a sum of 7 with 3 and 4, which leaves 1 and 2 for the fraction.

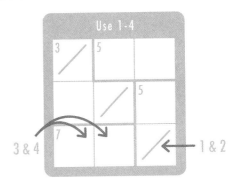

Since the 5 cage only has one fraction, that fraction must equal a whole number. So, we can only place 2/1 in the bottom-right corner.

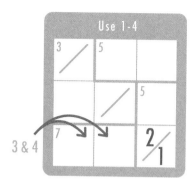

6. Find fractions with limited possibilities.

Sometimes it helps to know which digits are needed for a fraction even if we don't know which is the numerator and which is the denominator.

Which digits fill the highlighted squares?

The left column still needs a 1 and a 3, so the fraction in the left column is 1/3 or 3/1.

The bottom row still needs a 1 and a 2, so the fraction in the bottom row is 1/2 or 2/1.

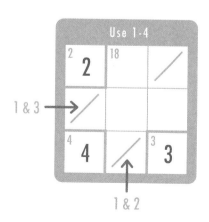

Then, which digit can fill the center square?

7. Make sure the sum of the squares in a cage isn't too small or too large.

What fraction fills the top-right corner?

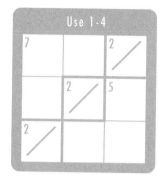

The top-right corner is either 2/1 or 4/2.

If we place 2/1 in the top-right corner, we must place 3 and 4 in the 5 cage, which makes the sum of the 5 cage too large.

So, we must place 4/2 in the top-right corner.

8. Find cages with unusually large or small sums.

Sometimes there are only a few ways to create a large or a small sum.

Beast Academy Puzzles 4

What digits fill the 11 cage?

We can only get a sum of 11 with 4+4+3.

There is only one way to place those digits without having two 4's in the same row or column.

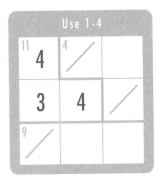

9. Fractions cannot equal 1.

A fraction equals 1 if it has the same numerator and denominator. We can't use the same digit twice in a row or column, so we cannot create fractions that equal 1.

What digit fills the top-left square?

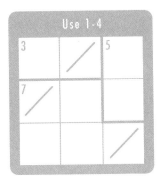

Since the sum in the top-left cage is 3, each square equals 1 or 2. Since the fraction can't equal 1, it must equal 2.

So, the other square in the 3 cage is 1.

(Then, what must the fraction be?)

10. Choose denominators so fractions sum to whole numbers.

What fraction fills the top-left corner?

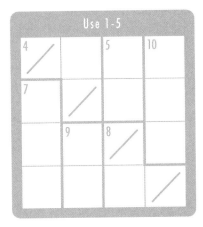

The 7 cage can only be filled with 1+2+4=7. The 9 cage can only be filled with 4+5=9.

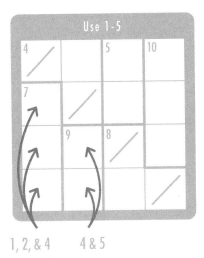

1, 2, & 4   4 & 5

So, the fraction in the left column is either 3/5 or 5/3.

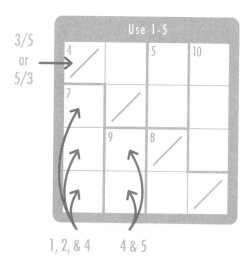

To give a whole number sum, the remaining fraction in the 4 cage must also be thirds or fifths. Since we already use a 5 in its column, both fractions must be thirds.

(Be careful with equivalent fractions – sometimes sixths equal thirds or halves.)

SUM SQUARES
DIFFICULTY LEVEL:
★—★★★★★

In a **Sum Square** puzzle, the digits 1 through 9 are used to fill the nine squares in the grid, one digit per square.

Some of the numbers in the grid are positive, and some are negative.

The numbers above and to the left of the grid give the sum of the integers in each column and row. On the right is an example of a completed Sum Square.

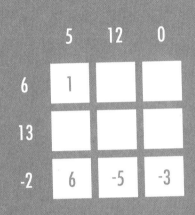

First, we look at the left column. We have 1+☐+6=5. This simplifies to 7+☐=5. Since 7+ -2 =5, we fill the middle square of the left column with -2 as shown.

The remaining digits are 4, 7, 8, and 9.

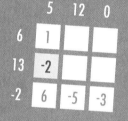

The two missing entries in the middle row must sum to 15 because -2+15=13. The only way to get a sum of 15 from two of the remaining digits is 7+8=15.

Similarly, the two missing entries in the middle column must sum to 17 because -5+17=12. The only way to get a sum of 17 from two of the remaining digits is 8+9=17.

We learned above that the 8 is in the middle row. So, we place the 8 as shown in the center square, with the 9 above it.

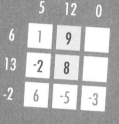

The remaining digits are 4 and 7.

In the top row, we have 1+9+ -4 =6, and in the middle row, we have -2+8+ 7 =13.

All the digits have now been placed, and we check the sum of the integers in each row and column.

1.

	7	5	3
11		3	7
-13	2		
17			5

2.

	1	2	4
		3	5
-5			6
0	-1		

3.

	2	9	0
4			-5
7		6	
0	3		-4

4.

	10	12	-15
9		8	-4
-6	6		-9
4			

5.

	4	5	6
2	3		
1		7	-8
			9

6.

	24	5	0
20			
9		2	-1
0		-3	-4

7.

	18	-5	18
9			1
13		-3	
9	5		8

8.

	1	11	21
10			9
11	-2		
12	-1	8	

9.

	5	2	2
6	2		3
6			
-3	-6		-5

10.

	-1	0	4
2	-6		3
-2		2	
3	1		

11.

	6	-7	
4		-6	9
8	7		
9			8

12.

		1	3
2			9
4	5		
7	8	6	

13.

	1	1	3
3			1
3			
-1	3		-6

14.

	15	15	-15
5			
5		1	-5
5		6	

15.

	-1	15	-9
3		9	
15			3
-13		1	

16.

	-2	4	13
3		1	
14		-2	
-2	-3		-4

17.

	5	5	5
6			1
5			
4	-7		9

18.

	1	3	5
5		-5	3
3		-1	
1			

★ ★ ☆ ☆ ☆

19.

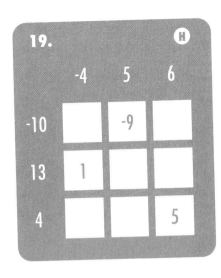

	-4	5	6
-10		-9	
13	1		
4			5

20.

	8		4
2		-4	
9	6		8
-6		2	

21.

	10	5	10
20	3		
5			
0			-6

22.

	10	-10	11
10			8
10		-3	
-9	-6		

23.

	9	11	-13
5	7		
3		5	
-1		9	

24.

	1	1	1
1			
1		-1	
1		-5	9

25.

	5	6	
4			7
3		-1	
	2		-8

26.

	-2	2	-21
-1	2	1	
1			
-21			

27.

	15	-15	15
1			
15		-1	
-1			5

28.

	1	0	0
1	2		
-10		6	
10			

29.

	2	1	
4			
	-8		-1
		4	2

30.

	4	-10	7
7			3
-10			
4			-2

Hints on pages 183-184

31.

32.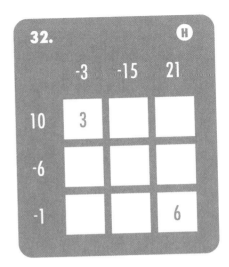

33.

34.

35.

36.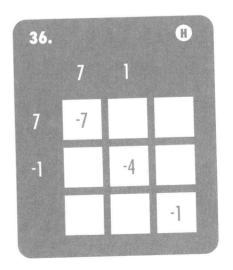

37.

38.

39.

40.

41.

42.

Ⓗ Hints on pages 183-184

SUM SQUARES
STRATEGIES

1. Write the unused digits above the puzzle.

Each Sum Square needs the digits 1 through 9. We can write the unused digits, then cross them off as we place them.

2. Complete rows or columns with one empty square.

If we know two numbers and the sum of a row or column, then we can find the third number.

Which number can we place first?

In the top row, we have 3 and 5, but we do not know the target sum, so we cannot complete that row.

In the right column, we have 5+6+☐=4. This simplifies to 11+☐=4. Since 11+(-7)=4, we place -7 in the bottom-right square.

3. Find all possible pairs of digits that can go in a row or column.

We can find the amount still needed to get a target sum, then write all the ways we can create that amount with the unused digits.

Which pairs of digits could go in each column?

The unused digits are 2, 4, 6, and 8. In the left column, we need to add a total of -6 to the 7 to get the target sum of 1.

If we choose two numbers with the same sign, we can only add -6 by using (-4)+(-2).

If we choose two numbers with different signs, we can only add -6 by using (-8)+2.

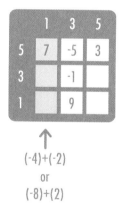

(-4)+(-2)
or
(-8)+(2)

So, we must use the digit 2 in the left column.

In the right column, we need to add a total of 2 to the 3 to get the target sum of 5. The unused digits are 4, 6, and 8.

Beast Academy Puzzles 4

We can only add 2 by using 6+(-4) or 8+(-6).

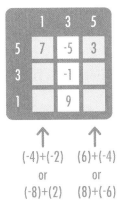

↑ ↑
(-4)+(-2) (6)+(-4)
or   or
(-8)+(2) (8)+(-6)

So, we must use the digit 6 in the right column.

4. Check intersections.

What number goes in the center square?

The unused digits are 4, 5, 7, and 9.

In the middle row, to get the target sum of 3 we must add a total of -5 to the 8. We can only do this by using 4 and -9.

In the middle column, to get the target sum of 1 we must add a total of -1 to the 2. We can only do this by using 4 and -5.

So, only 4 can be placed in the center square.

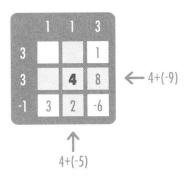

← 4+(-9)

↑
4+(-5)

5. Look for extreme sums.

If we need to add a lot or subtract a lot to reach a target sum, this limits our choices.

How can we fill the center-right square?

In the middle row, to get a sum of 14 we must add a total of 16 to the -2. We can only do this by placing 7 and 9 in the middle row.

In the right column, to get a sum of 13 we must add a total of 17 to the -4. We can only do this by placing 8 and 9 in the right column.

So, we must place 9 in the center-right square.

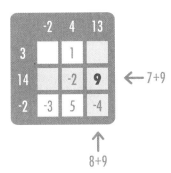

← 7+9

↑
8+9

6. Find regions where we can't place a given digit.

Sometimes placing a digit in a square would make it impossible to complete a row or column.

How can we complete the right column?

The unused digits are 1, 2, 3, 4, and 9.

In the right column, we need to add a total of 10 to the -7 to get the target sum of 3. We can only do this by placing 1 and 9 in the right column.

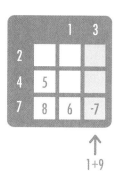

1+9

If we place the 9 in the middle row, there is no digit we can add to 5+9=14 to get the target sum of 4.

So, we place the 9 in the top-right square, and then place the 1 in the remaining blank of the right column.

7. Is there only one row or column where a number can go?

If a number can't be used in one row, then it can only be used in the other rows.

In which row do we place the 3?

The unused digits are 1, 3, 7, and 9.

In the bottom row, we need to add a total of -8 to the 2 to get the target sum of -6. We can do that with -1+(-7) or (-9)+1.

Since we can't place the 3 in the bottom row, we must place the 3 in the top row.

8. A number in a row cannot equal the row's sum.

If one number in a row equals that row's sum, then the other two numbers in that row must sum to 0.

But we can't make a sum of 0 from two different digits. So, we can't place a number in a row that is equal to the sum of the row.

How can we complete the middle column?

The unused digits are 3, 4, 5, 6, and 9.

In the middle column, to get the target sum of 6 we must add a total of 7 to the -1. We can only do this by placing 3 and 4 in the middle column.

We can't place a 4 in a row with a sum of 4, so we must instead place 4 in the bottom-middle square and 3 in the top-middle square.

(Similarly, the sum of two numbers in a row cannot equal the row's sum.)

9. Check the math.

It's easy to make a sign mistake. Does each completed row and column give the correct sum? Do the possible choices for completing a row and column all give the correct sum?

PAINT THE TOWN
DIFFICULTY LEVEL:
★—★★★★★

In a **Paint the Town** puzzle, the goal is to paint some of the houses in a grid.

A clue in each house gives the fraction of houses in its neighborhood that must be painted.

A house's neighborhood includes itself and all of the houses that surround it, including diagonally.

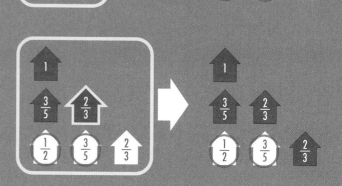

The $1=\frac{3}{3}$ clue tells us that 3 of the 3 houses in its neighborhood must be painted.

We 'paint' all three houses by filling them in.

The $\frac{3}{5}$ clue tells us that 3 of the 5 houses in its neighborhood must be painted. Since 3 of the 5 cells are already painted, we circle $\frac{1}{2}$ and $\frac{3}{5}$ to remind us not to paint these.

Finally, the $\frac{2}{3}=\frac{4}{6}$ clue tells us that 4 of the 6 cells in its neighborhood must be painted. So, we must paint the $\frac{2}{3}$ house in the bottom-right.

1.

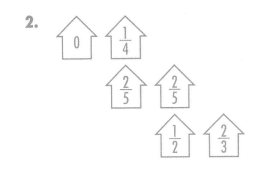

2.

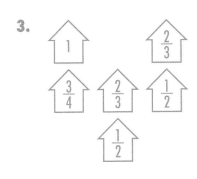

3.

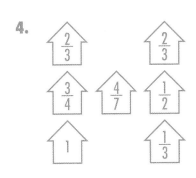

4.

5.

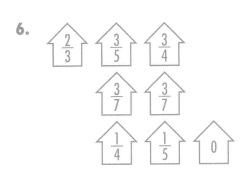

6.

7.

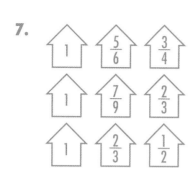

8.

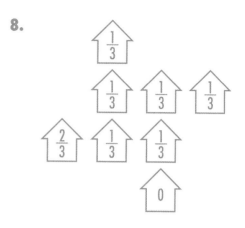

9.

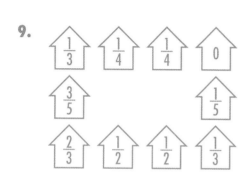

10.

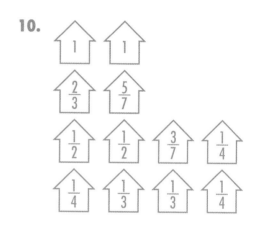

11.

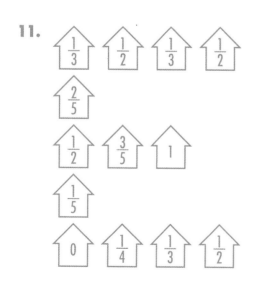

12.

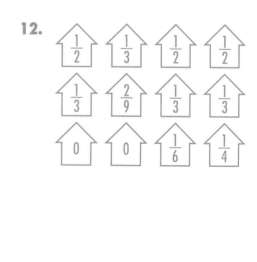

13.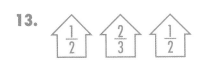

14.

15.

$$\frac{1}{4}$$

$$\frac{1}{2} \quad \frac{2}{5} \quad \frac{1}{4}$$

$$\frac{1}{2}$$

16.

$$\frac{1}{2}$$

$$\frac{1}{3}$$

$$\frac{1}{2}$$

$$\frac{1}{3} \quad \frac{1}{4} \quad \frac{2}{3} \quad \frac{1}{2}$$

17.

$$\frac{3}{4} \quad \frac{3}{5}$$

$$\frac{3}{5} \quad \frac{4}{7} \quad \frac{3}{5}$$

$$\frac{3}{5} \quad \frac{1}{2}$$

18.

$$\frac{1}{5} \quad \frac{2}{5}$$

$$\frac{1}{4} \quad \frac{1}{7} \quad \frac{2}{7} \quad \frac{1}{4}$$

$$\frac{1}{5} \quad \frac{2}{5}$$

19. H

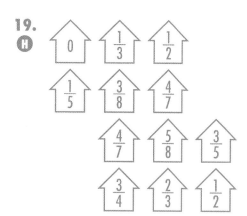

20. H

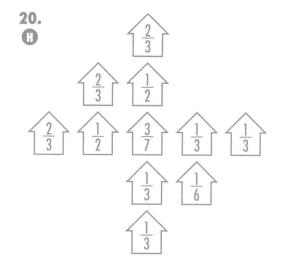

21. H

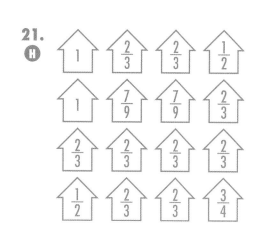

22. H

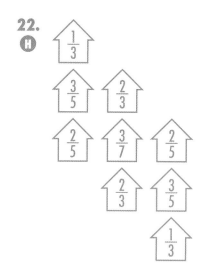

23. H

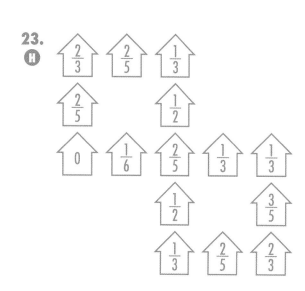

24. H

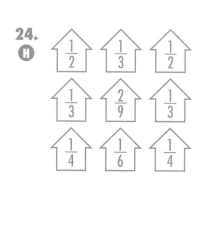

☆ ☆
★ ☆ ☆
★ ★
★

25.

26.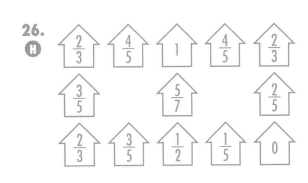

27.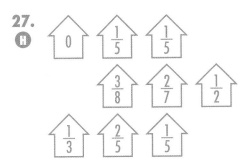

28.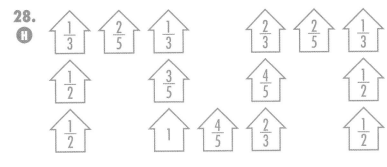

29.
H

30.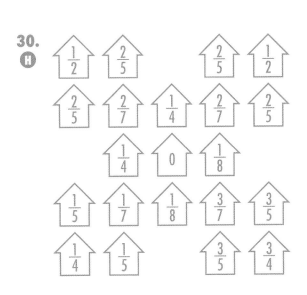

H Hints on pages 185-187

☆
☆
★
★
★

31. Ⓗ

$\frac{3}{4}$ $\frac{1}{2}$ $\frac{1}{4}$

$\frac{2}{3}$ $\frac{4}{9}$ $\frac{1}{3}$

$\frac{1}{2}$ $\frac{1}{3}$ $\frac{1}{4}$

32. Ⓗ

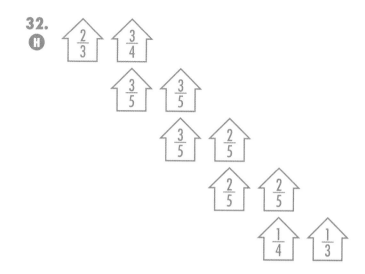

33. Ⓗ

$\frac{1}{3}$ $\frac{2}{5}$ $\frac{1}{2}$ $\frac{4}{5}$ 1

$\frac{2}{5}$ $\frac{3}{7}$ $\frac{3}{5}$

$\frac{1}{4}$ $\frac{2}{7}$ $\frac{1}{5}$ $\frac{3}{7}$ $\frac{1}{2}$

$\frac{2}{5}$ $\frac{2}{7}$ $\frac{1}{5}$

$\frac{1}{3}$ $\frac{1}{5}$ $\frac{1}{4}$ $\frac{1}{5}$ $\frac{1}{3}$

34. Ⓗ

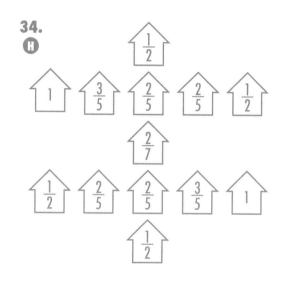

35. Ⓗ

$\frac{3}{4}$ $\frac{3}{5}$ $\frac{4}{5}$ $\frac{3}{4}$

$\frac{2}{3}$ $\frac{1}{2}$ $\frac{2}{7}$ $\frac{3}{7}$ $\frac{5}{8}$ $\frac{2}{3}$

$\frac{3}{4}$ $\frac{3}{7}$ $\frac{1}{4}$ $\frac{1}{4}$ $\frac{3}{7}$ $\frac{1}{2}$

0 $\frac{1}{5}$

36. Ⓗ

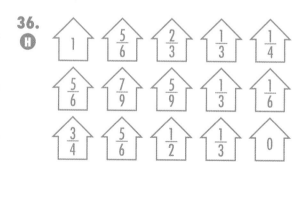

1 $\frac{5}{6}$ $\frac{2}{3}$ $\frac{1}{3}$ $\frac{1}{4}$

$\frac{5}{6}$ $\frac{7}{9}$ $\frac{5}{9}$ $\frac{1}{3}$ $\frac{1}{6}$

$\frac{3}{4}$ $\frac{5}{6}$ $\frac{1}{2}$ $\frac{1}{3}$ 0

37.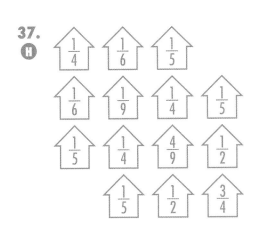

38.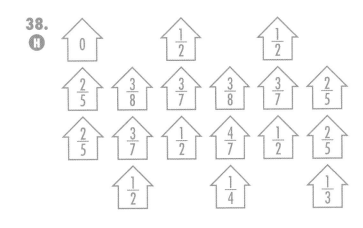

39.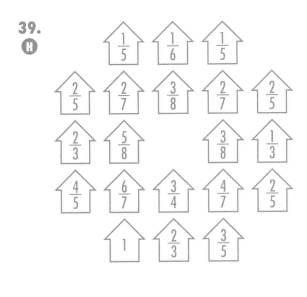

40.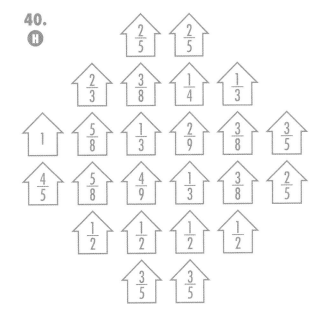

41.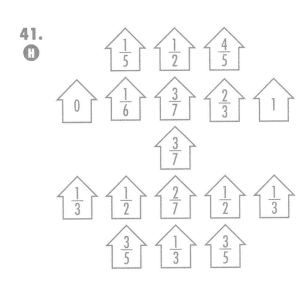

42.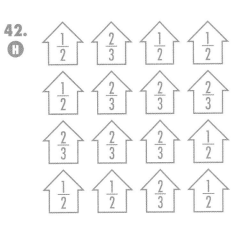

Hints on pages 185-187

43.

44.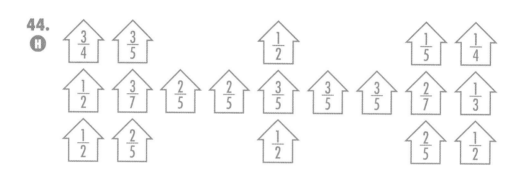

45.

❗ Hints on pages 185-187

46.

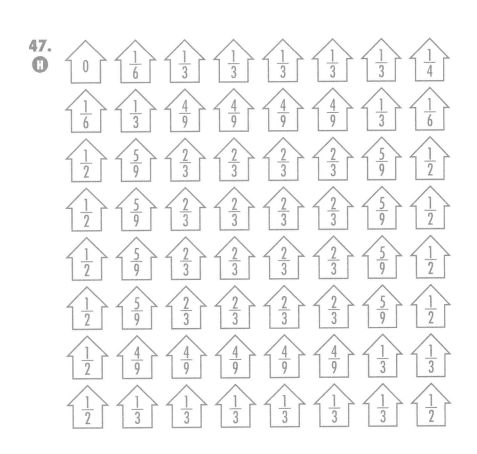

47.

Hints on pages 185-187

PAINT THE TOWN
STRATEGIES

1. Circle unshaded houses.

If we know a house must remain unshaded, we can mark this by circling that house.

2. Begin with 1's and 0's.

Any 1 has its entire neighborhood shaded.
Any 0 has its entire neighborhood unshaded.

How can we start the puzzle below?

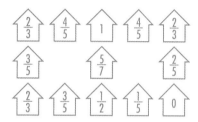

We shade all the houses in the neighborhood of the 1.

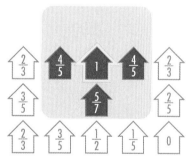

We circle all the houses in the neighborhood of the 0 to mark them as unshaded.

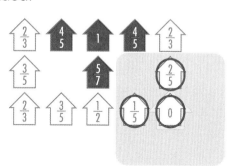

3. Find neighborhoods where all the shaded houses or all the unshaded houses are known.

Is the highlighted house shaded or unshaded?

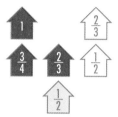

The $\frac{3}{4}$ clue tells us that 3 of the 4 houses in its neighborhood are shaded and the other 1 house is unshaded.

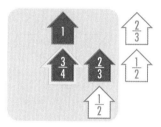

Since we have already shaded 3 houses in its neighborhood, the remaining house must be unshaded.

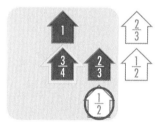

4. Mark groups of houses where we know only one is shaded.

If a pair of houses has exactly one shaded house, connect them to remember this.

Which houses in the neighborhood of the $\frac{3}{8}$ are shaded?

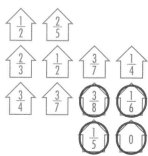

The $\frac{1}{6}$ tells us that 1 of the 2 unmarked houses in its neighborhood is shaded, so we link them.

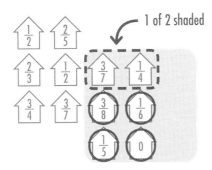

1 of 2 shaded

To shade 3 of the 8 houses in the neighborhood of the $\frac{3}{8}$, we must shade the two houses as shown, as well as one of the linked houses.

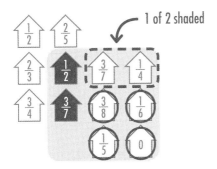

1 of 2 shaded

5. Compare neighborhoods.

If one neighborhood is inside another neighborhood, we can look at the difference between these two neighborhoods.

How do the neighborhoods of the two highlighted houses compare?

Only 2 of the 4 houses in the neighborhood of the bottom-right house are shaded.

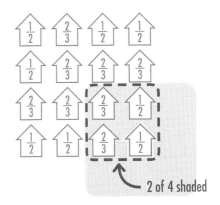

2 of 4 shaded

4 of the 6 houses in the neighborhood of the $\frac{2}{3}$ in the bottom row are shaded.

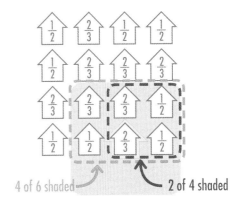

4 of 6 shaded   2 of 4 shaded

But since we know only two of the rightmost houses in that neighborhood are shaded, we must shade the other 2 houses to have 4 shaded houses in that neighborhood.

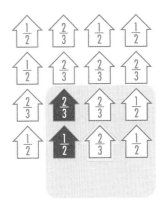

6. Combine clues.

What can we tell from the $\frac{3}{8}$ clue?

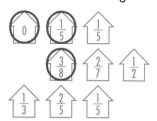

One house in the neighborhood of the circled $\frac{1}{5}$ is shaded.

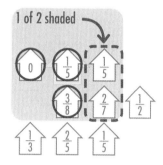

One house in the neighborhood of the $\frac{1}{3}$ is shaded.

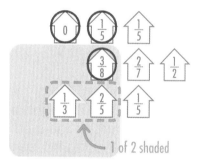

We can only shade three houses in the neighborhood of the $\frac{3}{8}$ by shading the $\frac{1}{5}$ in the bottom row.

So, we shade that house.

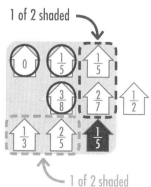

7. Keep looking!

Sometimes the real challenge is finding the one easiest clue, which may be far away from the clue we just used.

DECIMAL NUMBERCROSS
DIFFICULTY LEVEL:

In a **Decimal Numbercross** puzzle, the goal is to write all of the given fractions as decimals in the grid.

Each box will contain either a digit or a decimal point.

In these puzzles, a decimal can be written with any number of trailing 0's. For example, $\frac{32}{10}$ can be written as 3.2, or 3.20, or even 3.2000 if there is enough room.

A number less than 1 may have a 0 before the decimal point. For example, $\frac{2}{10}$ can be written as 0.2 or simply .2.

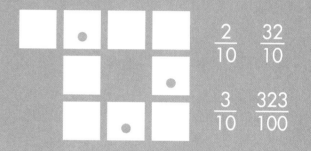

First, we write each of the fractions as decimals.

$\frac{2}{10}$ =.2, $\frac{32}{10}$ =3.2, $\frac{3}{10}$ =.3, and $\frac{323}{100}$ =3.23.

There is only one place in the grid where 3.23 fits. We write 3.23 in the top row as shown.

Then, we have 3.■ in the right column. Only 3.2 fits in the right column.

We add a 0 in the ones place to write 0.2 in the bottom row. Finally, we add a trailing zero to put .30 in the remaining space.

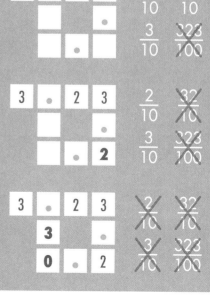

$$\frac{37}{10}$$

$$\frac{737}{10}$$

$$\frac{703}{1,000}$$

$$\frac{65}{10} \qquad \frac{56}{100}$$

$$\frac{605}{10} \qquad \frac{506}{1,000}$$

$$\frac{4}{10} \qquad \frac{444}{10}$$

$$\frac{44}{10} \qquad \frac{4}{100}$$

$$\frac{22}{10} \qquad \frac{22}{100}$$

$$\frac{52}{10} \qquad \frac{405}{100}$$

$$\frac{7}{10} \qquad \frac{2}{100}$$

$$\frac{25}{10} \qquad \frac{75}{100}$$

$$\frac{3}{10} \qquad \frac{33}{100}$$

$$\frac{35}{10} \qquad \frac{77}{100}$$

$$\frac{75}{10}$$

7.

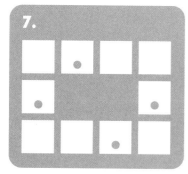

$\dfrac{2}{10}$ $\dfrac{8}{10}$

$\dfrac{4}{10}$ $\dfrac{248}{10}$

8.

$\dfrac{4}{10}$ $\dfrac{22}{100}$

$\dfrac{11}{100}$ $\dfrac{144}{100}$

9.

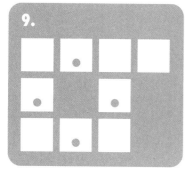

$\dfrac{1}{10}$ $\dfrac{81}{10}$

$\dfrac{8}{10}$ $\dfrac{7}{100}$

10.

$\dfrac{2}{10}$ $\dfrac{22}{100}$

$\dfrac{5}{10}$ $\dfrac{252}{100}$

11.

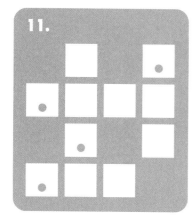

$\dfrac{1}{10}$ $\dfrac{33}{100}$

$\dfrac{111}{10}$ $\dfrac{103}{1,000}$

12.

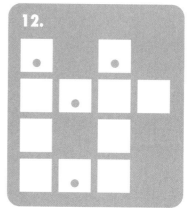

$\dfrac{2}{10}$ $\dfrac{272}{100}$

$\dfrac{9}{10}$ $\dfrac{729}{1,000}$

13.

$$\frac{404}{10} \qquad \frac{404}{100}$$

$$\frac{4}{100} \qquad \frac{404}{1,000}$$

14.

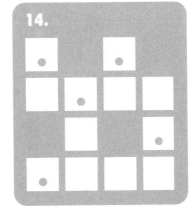

$$\frac{2}{10} \qquad \frac{5}{10}$$

$$\frac{3}{10} \qquad \frac{24}{10}$$

$$\frac{4}{10} \qquad \frac{5}{1,000}$$

15.

$$\frac{789}{10} \qquad \frac{567}{1,000}$$

$$\frac{67}{100} \qquad \frac{789}{1,000}$$

16.

$$\frac{1}{10} \qquad \frac{54}{10}$$

$$\frac{5}{10} \qquad \frac{3}{100}$$

$$\frac{6}{10} \qquad \frac{632}{100}$$

17.

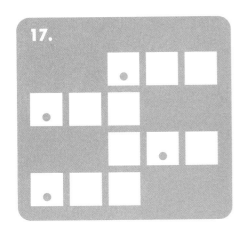

$$\frac{6}{10} \qquad \frac{24}{100}$$

$$\frac{48}{10} \qquad \frac{36}{100}$$

$$\frac{12}{100}$$

18.

$$\frac{505}{10}$$

$$\frac{205}{100}$$

$$\frac{208}{100}$$

$$\frac{805}{100}$$

19.

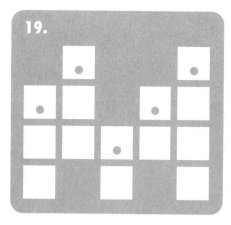

$$\frac{1}{10} \qquad \frac{4}{10}$$

$$\frac{4,444}{100}$$

$$\frac{2}{10} \qquad \frac{34}{100}$$

$$\frac{246}{1,000}$$

$$\frac{3}{10} \qquad \frac{43}{100}$$

20.

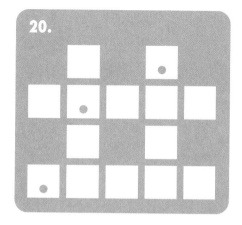

$$\frac{5}{10} \qquad \frac{55}{100}$$

$$\frac{5}{100} \qquad \frac{5}{10,000}$$

21.

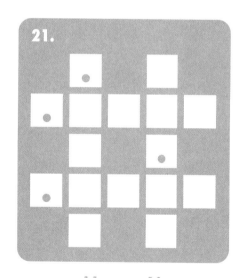

$$\frac{11}{100} \qquad \frac{11}{1,000}$$

$$\frac{1,001}{100} \qquad \frac{101}{10,000}$$

22.

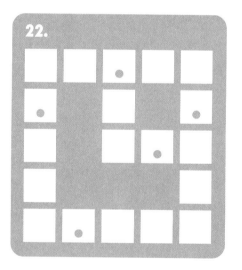

$$\frac{4}{10} \qquad \frac{14}{100}$$

$$\frac{44}{10} \qquad \frac{22}{100}$$

$$\frac{505}{10} \qquad \frac{555}{100}$$

23.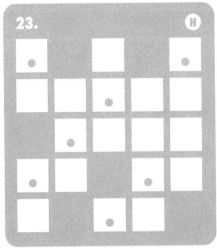

$$\frac{3}{10} \qquad \frac{6}{10} \qquad \frac{57}{10}$$

$$\frac{345}{1,000}$$

$$\frac{4}{10} \qquad \frac{7}{10} \qquad \frac{506}{10}$$

$$\frac{804}{1,000}$$

$$\frac{5}{10} \qquad \frac{8}{10} \qquad \frac{3,553}{100}$$

24.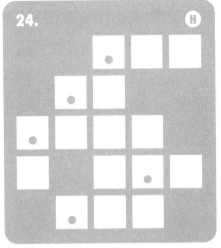

$$\frac{1}{10} \qquad \frac{9}{10} \qquad \frac{93}{100}$$

$$\frac{3}{10} \qquad \frac{93}{10} \qquad \frac{97}{100}$$

$$\frac{5}{10} \qquad \frac{19}{100} \qquad \frac{9,999}{10,000}$$

25.

$$\frac{2,002}{10} \qquad \frac{2,002}{1,000}$$

$$\frac{23}{100} \qquad \frac{2,132}{1,000}$$

$$\frac{1,203}{1,000} \qquad \frac{2,312}{1,000}$$

26.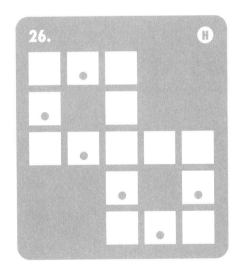

$$\frac{1}{10} \qquad \frac{52}{10}$$

$$\frac{2}{10} \qquad \frac{55}{10}$$

$$\frac{5}{10} \qquad \frac{1,055}{10}$$

Hints on page 188

27.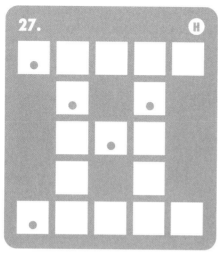

$$\frac{2}{10}$$

$$\frac{22}{100}$$

$$\frac{33}{1,000}$$

$$\frac{33}{100}$$

$$\frac{1}{10,000}$$

28.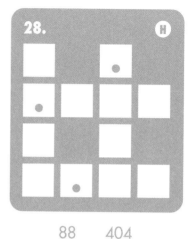

$$\frac{88}{100} \qquad \frac{404}{100}$$

$$\frac{234}{100} \qquad \frac{789}{1,000}$$

29.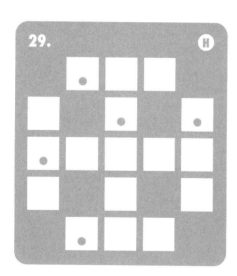

$$\frac{2}{10} \qquad \frac{1}{100}$$

$$\frac{6}{10} \qquad \frac{66}{100}$$

$$\frac{66}{10} \qquad \frac{2,006}{1,000}$$

30.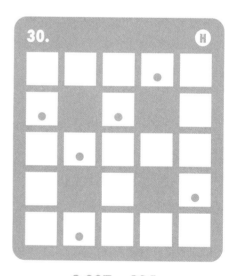

$$\frac{2,007}{10} \qquad \frac{205}{100}$$

$$\frac{7,002}{10} \qquad \frac{3}{1,000}$$

$$\frac{6}{100} \qquad \frac{302}{1,000}$$

31.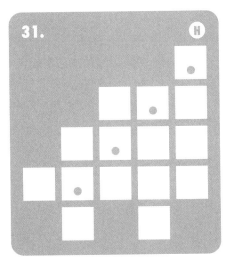

$$\frac{4}{10} \quad \frac{45}{100}$$

$$\frac{5}{10} \quad \frac{54}{100} \quad \frac{444}{1,000}$$

$$\frac{6}{10} \quad \frac{65}{100}$$

32.

$$\frac{233}{100}$$

$$\frac{222}{10} \quad \frac{22}{100}$$

$$\frac{322}{100}$$

$$\frac{223}{10} \quad \frac{33}{100}$$

$$\frac{522}{100}$$

33.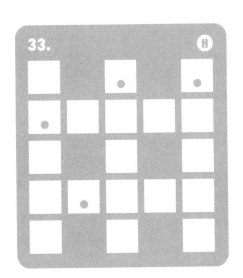

$$\frac{8}{100}$$

$$\frac{808}{1,000}$$

$$\frac{88}{100}$$

$$\frac{888}{1,000}$$

$$\frac{88}{1,000}$$

34.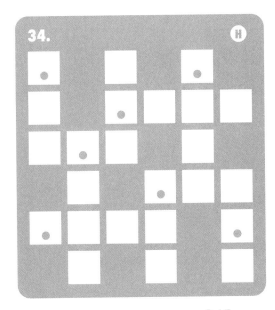

$$\frac{4}{10}$$

$$\frac{345}{1,000}$$

$$\frac{43}{10} \quad \frac{3}{100}$$

$$\frac{5}{10}$$

$$\frac{876}{1,000}$$

$$\frac{45}{10} \quad \frac{54}{100}$$

$$\frac{6}{10}$$

$$\frac{789}{1,000}$$

H Hints on page 188

These puzzles don't show the decimal points.

You'll need to place them on your own.

35. (H)

$\dfrac{6}{10}$ $\dfrac{43}{100}$

$\dfrac{25}{100}$ $\dfrac{306}{100}$

36. (H)

$\dfrac{3}{10}$ $\dfrac{31}{100}$

$\dfrac{17}{10}$ $\dfrac{95}{100}$

$\dfrac{57}{10}$

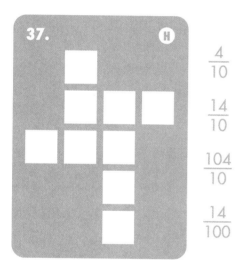

37. (H)

$\dfrac{4}{10}$

$\dfrac{14}{10}$

$\dfrac{104}{10}$

$\dfrac{14}{100}$

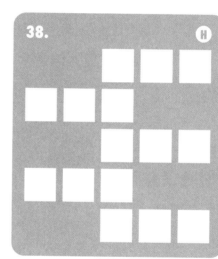

38. (H)

$\dfrac{12}{10}$ $\dfrac{23}{100}$

$\dfrac{32}{10}$ $\dfrac{34}{100}$

$\dfrac{43}{10}$ $\dfrac{1,234}{100}$

39. (H)

$\dfrac{12}{10}$

$\dfrac{123}{100}$

$\dfrac{12}{1,000}$

$\dfrac{123}{10,000}$

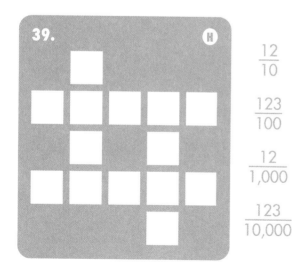

40. (H)

$\dfrac{1}{10}$ $\dfrac{12}{10}$

$\dfrac{2}{10}$ $\dfrac{34}{10}$

$\dfrac{3}{10}$ $\dfrac{1}{100}$

$\dfrac{4}{10}$

41.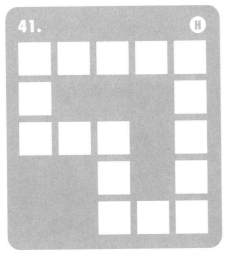

$$\frac{13}{10} \qquad \frac{51}{10}$$

$$\frac{33}{10} \qquad \frac{55}{10}$$

$$\frac{37}{10} \qquad \frac{777}{1{,}000}$$

42.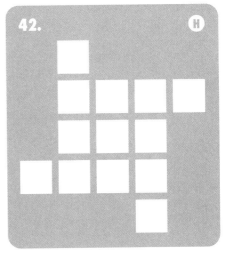

$$\frac{2}{10} \qquad \frac{501}{10}$$

$$\frac{3}{10} \qquad \frac{3}{100}$$

$$\frac{42}{10} \qquad \frac{54}{100}$$

43.

$$\frac{1}{10} \qquad \frac{212}{10}$$

$$\frac{12}{10} \qquad \frac{1}{100}$$

$$\frac{121}{10} \qquad \frac{212}{100}$$

44.

$$\frac{77}{10}$$

$$\frac{77}{100} \qquad \frac{77}{1{,}000}$$

$$\frac{707}{100} \qquad \frac{77}{10{,}000}$$

H Hints on page 188

DECIMAL NUMBERCROSS

STRATEGIES

1. Write the decimals.

Convert the fractions into decimals.

Write the decimals with their place values aligned. This will make it easier to find digits in specific place values.

$$
\begin{array}{r}
40.4 \\
4.04 \\
.04 \\
.404
\end{array}
$$

(Remember, in these puzzles, we can write a decimal with any number of trailing 0's. A decimal less than 1 can have a 0 before the decimal point, but does not have to.)

2. Cross out numbers as you use them.

After you write a decimal in the grid, cross out the number you used to remind you not to try to use it again.

3. Find numbers that only fit in one place.

Sometimes a number has an unusually large number of digits before or after the decimal point, limiting the places where it can be written.

Which number can only go in one place?

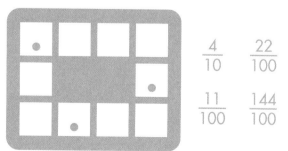

$\frac{4}{10}$ $\frac{22}{100}$

$\frac{11}{100}$ $\frac{144}{100}$

$\frac{4}{10}$ can be written as 0.4, .40, 0.40, or .400. So, it can go in any of the four places.

$\frac{11}{100}$ can be written as .11, 0.11, or 0.110, so it can go in three of the four places.

Similarly, $\frac{22}{100}$ can go in three of the four places.

$\frac{144}{100}=1.44$ has one digit to the left of the decimal point and at least two digits to the right of the decimal point. There is only one place it can go.

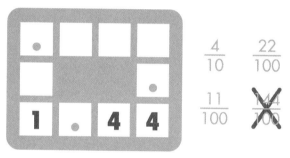

$\frac{4}{10}$ $\frac{22}{100}$

$\frac{11}{100}$ ✗$\frac{144}{100}$

(This strategy is best for numbers with lots of digits to the right of the decimal point, or at least one digit to the left of the decimal point.)

4. Find places where only one number fits.

Sometimes a place can only fit one number.

Which number goes in the highlighted row?

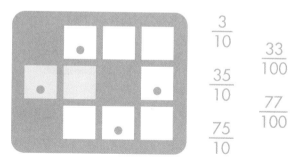

$\frac{3}{10}$ $\frac{33}{100}$

$\frac{35}{10}$ $\frac{77}{100}$

$\frac{75}{10}$

The decimal in the highlighted row can't have digits to the left of the decimal point, so it can't be $\frac{35}{10}=3.5$ or $\frac{75}{10}=7.5$.

It can have at most one digit to the right of the decimal point, so it can't be $\frac{33}{100}=.33$ or $\frac{77}{100}=.77$.

So, we can only put $\frac{3}{10}=.3$ in the middle row.

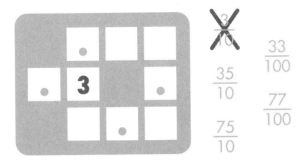

5. Eliminate choices with intersections.

Which digit can we write in the top-left corner?

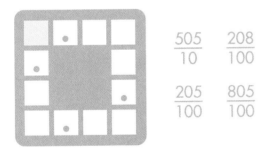

The digit in the top-left corner is the ones digit of two different numbers.

The decimals we have to place are $\frac{505}{10}=50.5$, $\frac{208}{100}=2.08$, $\frac{205}{100}=2.05$, and $\frac{805}{100}=8.05$.

2 is the only digit that appears in the ones place of two different numbers.

So, we write a 2 in the top-left corner.

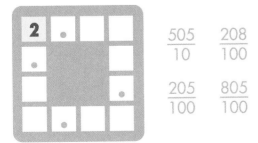

6. Look for common digits.

If all the remaining numbers have the same digit in a place value, we can write that digit in the correct place value for every number in the grid.

What place value has the same digit for all of the numbers below?

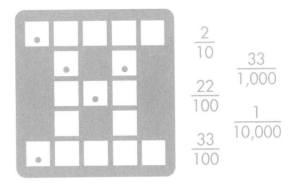

Each number is less than one, so the digit in every ones place is 0. We write a 0 in the ones place for every number in the grid.

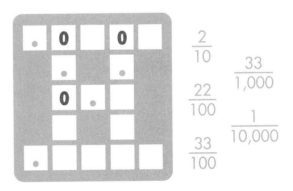

(Then, what number can we put in the top row?)

7. Where are the decimal points?

Some of the puzzles don't give the decimal points. But we can often use the numbers to locate the decimal points.

Where can we write the decimal points in the puzzle below?

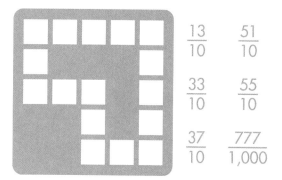

The decimals are $\frac{13}{10}$=1.3, $\frac{33}{10}$=3.3, $\frac{37}{10}$=3.7, $\frac{51}{10}$=5.1, $\frac{55}{10}$=5.5, and $\frac{777}{1,000}$=0.777.

We look at the places in the grid that are three boxes long. All the numbers that can fit in those places have a ones digit and a tenths digit, so the decimal point must go in the middle box.

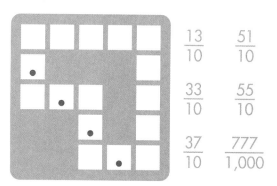

All the given numbers are less than 10. So, in the top row, the decimal point can only go in one of the two left-most boxes. If we put the decimal point in the top-left box, we would have two decimal points in the left column. So, we write the decimal point in the top row as shown.

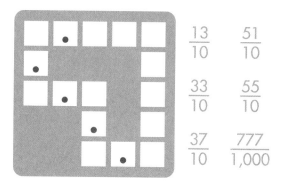

Similarly, the decimal point in the right column goes in the second box.

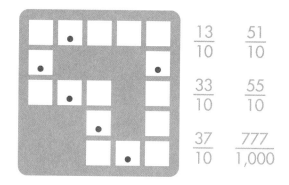

HINTS

DOT PUZZLES
(Strategies begin on page 16.)

13. When looking for a rhombus, look for diagonals that cross at a right angle and cut each other in half. (Strategy 10)

14. To make 2 parallelograms takes 2×4=8 dots. So, only 1 of the 9 dots will not be used. Are there any dots that can only be part of one parallelogram? (Strategies 3, 4)

15. There are three rectangles. Which of these do not share any corners? (Strategy 5)

16. There are two squares. Which of these does not share any corners with a rectangle? (Strategies 5, 8)

17. Since the dots are symmetrical, any shape we find in the top half will also appear in the bottom half. (Strategy 5)

18. Is there a rectangle that uses the blue dot below? (Strategy 4)

19. Find the square first. (Strategy 1)

20. There are 6 rectangles that are not squares. Look for rectangles in two different sizes. (Strategy 5)

21. There are four squares. Which of these do not share any corners? (Strategies 5, 8)

22. There are eight squares. Look for squares in different sizes that do not share any corners. (Strategies 5, 8)

23. Find the square first. (Strategy 1)

24. When looking for rhombuses, look for diagonals that cross at a right angle and cut each other in half. (Strategy 10)

25. Find the only rhombus that is not a square. Careful! Make sure it's a rhombus. (Strategies 7, 10)

26. Look for rectangles that have one corner in each group of three dots. Careful, the shape below is not a rectangle. Use strategy 9 to check for right angles.

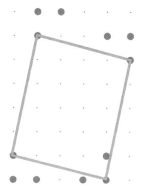

27. To make 3 squares takes 3×4=12 dots. So, only 1 of the 13 dots will not be used. Are there dots that can only be part of one square? Draw that square. Then, look for squares we can draw with the remaining dots. (Strategies 3, 4)

28. To make 3 squares takes 3×4=12 dots. So, only 2 of the 14 dots will not be used. Are there dots that can only be part of one square? Draw that square. Then, look for squares we can draw with the remaining dots. (Strategies 3, 4)

29. When looking for parallelograms, look for pairs of dots that can form opposite sides. (Strategy 11)

30. There are two squares. Which of these lets us draw a rectangle and a parallelogram with the remaining dots? (Strategies 1, 5)

31. To make 4 squares takes 4×4=16 dots. So, all 16 of the dots will be used. Are there any dots that can only be part of one square? If so, that square must be used. (Strategies 3, 4, 8)

32. To make 4 rectangles takes 4×4=16 dots. So, all 16 of the dots will be used. Are there any dots that can only be part of one rectangle? If so, that rectangle must be used. Look at the dots in the bottom-right corner. Which of these can only be part of one rectangle? (Strategies 3, 4)

33. Look for the shapes in the order they are listed, starting with the square. To make four quadrilaterals takes 4×4=16 dots. So, all 16 dots must be used. (Strategies 1, 3)

34. To make 5 squares takes 5×4=20 dots. So, all 20 dots will be used. Are there any dots that can only be part of one square? Draw that square. Then, look for squares we can draw with the remaining dots. (Strategies 3, 4)

35. How many squares are there? Can more than one square be part of the solution? (Strategies 1, 2, 5)

36. Find the squares first. Stay organized, checking each dot to see whether it can be part of a square. (Strategies 1, 5)

37. How can the dots that are not part of the groups in the top-left and bottom-right corners be used? Look for the rhombus first. There are two that are not squares. Careful! Make sure the shape is a rhombus. (Strategies 7, 10)

38. A rectangle cannot have more than 2 of its corners on the long diagonal line of dots. There are only 6 dots that are not on the diagonal, shown in blue below.

So, to draw 3 rectangles, every rectangle must use exactly 2 dots on the diagonal and 2 dots that are not on the diagonal. (In other words, no rectangle can use 3 of the blue dots above.)

How can we use the blue dot that is below the diagonal with exactly one of the other five blue dots to make a rectangle? There is more than one way (Strategy 9).

SPIRAL GALAXIES
(Strategies begin on page 30.)

15. Start with the three corners that contain galaxy centers. (Strategy 4)

16. Which galaxy contains the square in the top-left corner? (Strategy 7)

17. Which galaxy contains the square in the bottom-left corner? (Strategy 7)

18. Which galaxy contains the square in the bottom-right corner? (Strategy 7)

19. Which galaxy contains the square in the middle of the top row? (Strategy 7)

20. Which galaxy contains the square in the bottom-right corner? (Strategy 7)

21. Which galaxy contains the square in the middle of the bottom row? (Strategy 7)

22. Start at the corners and edges and work towards the middle. (Strategies 3, 4)

23. Which galaxy contains the highlighted square? (Strategy 7)

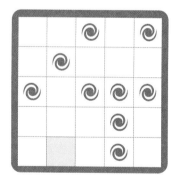

24. Which galaxy contains the square in the center? (Strategy 7)

25. Which galaxy contains the highlighted square? (Strategy 7)

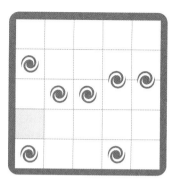

26. Which galaxy contains the square in the top-right corner? (Strategy 7)

27. Which galaxy contains the highlighted square? (Strategy 7)

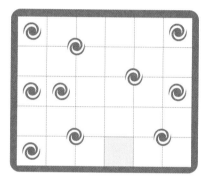

28. Which galaxy contains the square in the bottom-right corner? (Strategy 7)

29. Which galaxy contains the third square in the top row? Which galaxy contains the fourth square in the top row? (Strategy 7)

30. Which galaxy contains the highlighted square? (Strategy 7)

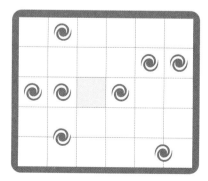

31. Which galaxy contains the square in the bottom-right corner? (Strategy 7)

32. Which galaxy contains the third square in the bottom row? Which galaxy contains the fourth square in the bottom row? (Strategy 7)

33. Which galaxy contains the square in the top-right corner? (Strategy 7)

34. Which galaxy contains the square in the bottom-left corner? (Strategy 7)

35. Which galaxy contains the square in the top-left corner? (Strategy 7)

36. Which galaxy contains the third square in the top row? (Strategy 7)

37. Which galaxy contains the square in the bottom-left corner? (Strategy 7)

38. Which galaxy contains the square in the top-left corner? (Strategy 7)

39. Which galaxy contains the square in the top-left corner? Which galaxy contains the highlighted square? (Strategy 7)

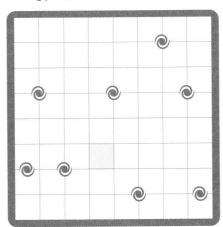

40. Which galaxies can contain the highlighted squares below? (Strategy 7)

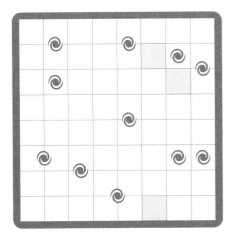

41. Which galaxies can contain the highlighted squares below? (Strategy 7)

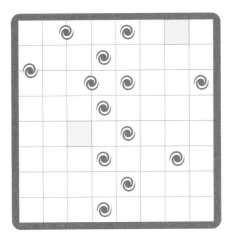

42. Which galaxy contains the square in the bottom-right corner? The bottom-left corner? (Strategy 7)

PRODUCT PLACEMENT
(Strategies begin on page 42.)

13. What is the second partial product? Then, what digit is a common factor of both partial products? (Strategies 5, 10)

14. What is the first partial product? Then, what digit is a common factor of both partial products? (Strategies 5, 10)

15. What is the ones digit of the first partial product? Then, what is the one-digit factor? (Strategies 6, 9)

16. What is the first partial product? (Strategy 8)

17. What is the second partial product? (Strategy 3)

18. What is the first partial product? (Strategy 9)

19. What is the ones digit of the first partial product? Then, what is the first partial product? (Strategies 6, 8)

20. What are the possibilities for the second partial product? Which of these can give us a three-digit final product? (Strategy 11)

21. What is the first partial product? (Strategy 8)

22. What is the tens digit of the first partial product? (Strategies 1, 6)

23. What are the ones digits of the two factors? (Strategy 7)

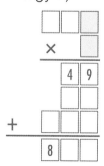

24. What is the first partial product? Then, what is the middle digit of the second partial product? (Strategies 3, 6)

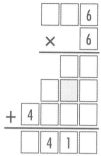

25. What is the second partial product? (Strategy 9)

26. What digits in the factors give us the third partial product? (Strategy 7)

27. The final product has 4 digits. What is the third partial product? (Strategy 11)

28. What are the possibilities for the second partial product? Which of these works? (Strategy 9)

29. What 1-digit factor can give the third partial product below and a 4-digit final product? (Strategy 11)

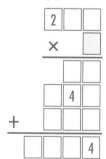

30. What is the tens digit of the first partial product?

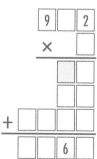

Then, what is the second partial product? (Strategy 6)

31. What is the middle digit of the second partial product?

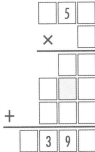

Then, what is the tens digit of the first partial product? (Strategy 6)

32. What is the third partial product? (Strategy 8)

33. What is the third partial product? (Strategies 6, 8)

34. The final product has 4 digits. What is the third partial product? (Strategies 6, 11)

35. What is the third partial product? (Strategies 6, 8, 10)

36. What are the possibilities for the first partial product? Then, is the middle digit of the second partial product odd or even?

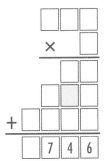

What does this tell us about the one-digit factor?

37. What is the fourth partial product? Then, what is the first partial product? (Strategies 6, 9, 10)

38. What are the possibilities for the third partial product that give us a 9 in the hundreds place of the final product? Which of these lets us get a 3 in the tens place of the final product? (Strategy 6)

39. What is the first partial product? Then, what is the fourth partial product that will let us create a 4-digit final product? (Strategies 6, 8, 11)

40. What digits can fill the highlighted box below?

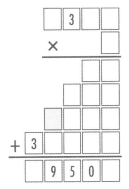

Then, what is the fourth partial product? (Strategy 6)

PYRAMID DESCENT
(Strategies begin on page 54.)

31. $3025=121\times25$. How could we complete the prime factorization of 3025 to help us solve the puzzle? (Strategy 1)

32. 30^4
$=(2\times3\times5)^4$
$=(2\times3\times5)\times(2\times3\times5)\times(2\times3\times5)\times(2\times3\times5)$.
How can we pair these factors to make 6's, 10's, and 15's? (Strategy 1)

33. $4864=304\times16$. How could we complete the prime factorization of 4864 to help us solve the puzzle? (Strategy 1)

34. $1056=132\times8$. How could we complete the prime factorization of 1056 to help us solve the puzzle? (Strategy 1)

35. 39^5
$=(3\times13)^5$
$=(3\times13)\times(3\times13)\times(3\times13)\times(3\times13)\times(3\times13)$
$=3^5\times13^5$.
(Strategy 1)

36. How could looking at the pyramid help us find the prime factorization of 1001? (Strategy 1)

37. How can we be sure to cross an odd number of 5's? (Strategy 4)

38. Every block contains a 1, 11, or 13. How many 1's can the path cross? (Strategy 3)

39. What are the possibilities for the total number of 5's we must cross in the bottom two rows? (Strategy 3)

40. $793,800=1,620\times490$. How could we complete the prime factorization of 793,800 to help us solve the puzzle? (Strategy 1)

DUTCH LOOP
(Strategies begin on page 64.)

21. How does the path pass through each ○? How can we draw half of the path through the ●? (Strategies 3, 7)

22. After drawing the partial paths through the corners, how does the path pass through the highlighted square below? (Strategies 2, 3)

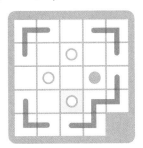

23. After drawing the partial paths through the corners, how can we extend the loose ends of the partial path in the top-right corner? (Strategies 2, 4)

24. After drawing the partial paths through the corners, how does the path pass through the highlighted square below? (Strategies 2, 7)

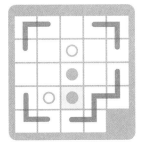

25. After drawing the partial paths through the corners, how does the path pass through each ○? (Strategy 3)

26. After drawing the partial paths through the corners, what parts of the path can we extend from the ●s in the highlighted squares below?

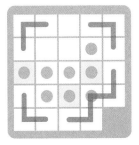

What walls can we draw to separate squares that cannot be connected by a path? (Strategies 2, 7, 9)

27. How must the path pass through the two ○s in the bottom row? Then, how must the path pass through the two ○s that are farthest right? (Strategies 3, 6)

28. What happens if the path through one of the ○s is horizontal? (Strategy 5)

29. After drawing the partial paths through the corners, how does the path pass through each of the ○s? (Strategies 2, 3, 6)

30. After drawing the partial paths through the corners and through the ○s with only one option, what parts of the path can we extend from the ●s in the highlighted squares below? (Strategies 2, 3, 7)

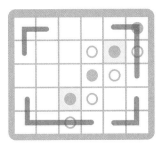

31. After drawing the partial paths through the corners, can we find squares that only have two open sides? (Strategies 2, 8)

32. After drawing the partial paths through the corners and through the ○s with only one option, what parts of the path can we extend from the ●s in the bottom row?

Then, how does the path pass through the highlighted squares? (Strategies 2, 3, 7, 9)

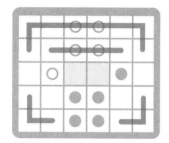

33. How does the path pass through the leftmost ○? How does it pass through the ○ in the second column? The third? Keep going. (Strategies 2, 3)

34. How does the path pass through the ○s in the bottom row? How does the path pass through the ○s in the third column? Then, how does the path pass through the ○s in the second row? (Strategies 3, 6)

35. How does the path pass through the eight squares below? (Strategies 2, 7)

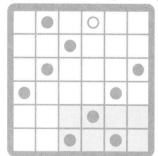

36. After we've drawn the path around the edge of the puzzle, what squares have only two open sides? (Strategies 2, 3, 8)

37. What happens if the path through one of the ○s in the second column is vertical? (Strategy 5)

38. How does the path pass through the ○s on the edges? Then, how does the path pass through the ○s that are not on the edges? (Strategy 3)

39. Focus on one part of the grid at a time. How can the path pass through the squares highlighted below? (Strategies 2, 3, 4, 8, 9)

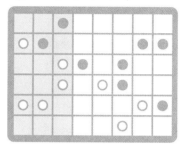

40. How will the path continue from the highlighted square below?

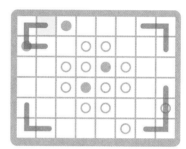

Then, how can we extend the loose ends in the top-left corner? (Strategy 5)

41. What happens if the path through one of the ○s in the fourth column is vertical? (Strategy 5)

42. What parts of the path can we extend in one direction from ●s in the grid? Then, draw walls between squares that cannot be connected by a path to help us draw more partial paths. (Strategies 7, 9)

HIVE
(Strategies begin on page 78.)

15. How can we fill the empty hexagons that touch the given 4? (Strategies 2, 3)

16. How can we fill the empty hexagons that touch the given 5? (Strategies 2, 3)

17. How can we fill the empty hexagon that touches the given 2? (Strategy 2)

18. How can we place a 1 that touches the given 2? (Strategy 3)

19. How can we place a 4 that touches the given 5? (Strategy 6)

20. How can we place the 2 that touches the 4 in the bottom row? Then, how can we place a 2 that touches the other given 4? (Strategy 3)

21. How can we place a 4 that touches the given 5? (Strategies 3, 6)

22. Where can we place the 1's? (Strategy 4)

23. What numbers must touch the highlighted 3? The 5? (Strategy 7)

24. How can we place the 1's? (Strategies 3, 4)

25. How can we place the 3 that touches the given 4? (Strategy 6)

26. How can we place a 3 that touches the given 4 so that the 3 can still touch a 2? (Strategies 3, 8)

27. How can we fill the empty hexagons that touch the given 4? (Strategy 3)

28. How can we place a 4 that touches the highlighted 5? (Strategy 6)

29. How can we place a 5 that touches the given 6? (Strategy 6)

30. How can we place a 4 that touches the highlighted 5? (Strategies 3, 6)

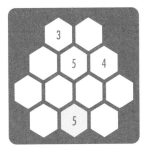

31. How can we place a 4 that touches the given 5? (Strategies 3, 6)

32. How can we place a 4 that touches the given 5? (Strategies 3, 6)

33. How can we fill the empty hexagons that touch the given 5? (Strategies 3, 6)

34. Where can we place the 3's? (Strategies 2, 6)

35. How can we fill the empty hexagons that touch the highlighted 5? (Strategies 3, 6)

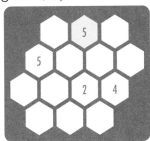

36. How can we fill the empty hexagons that touch the highlighted 5? (Strategies 3, 6)

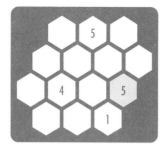

37. How can we fill the empty hexagons that touch the given 3? Then, the given 6? (Strategies 3, 6)

38. How can we place a 5 that touches the given 6? Then, how can we fill the empty hexagons that touch the given 3? (Strategies 3, 4, 6)

39. How can we fill the empty hexagons that touch the highlighted 5's? (Strategies 3, 6)

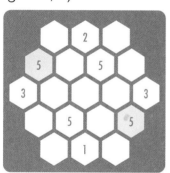

40. How can we fill the empty hexagons that touch the given 6? (Strategy 3)

41. What numbers must touch the highlighted 5? The 3? (Strategy 7)

42. How can we place a 3 that touches the highlighted 4? (Strategy 3)

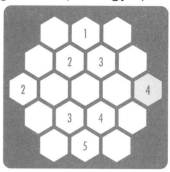

43. How can we place a 1 and 2 that touch the highlighted 3? How can we place a 4 that touches the highlighted 5? (Strategies 3, 6)

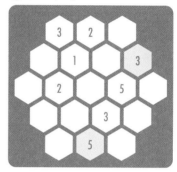

44. How can we place a 2 that touches the given 5? Then, which numbers can fill the empty hexagons that surround the given 5? The given 6? (Strategies 3, 7)

45. How can we fill the empty hexagons that touch the highlighted 5? Then, where can we place the 1's? (Strategies 3, 4, 6)

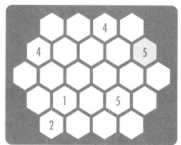

46. How can we place a 4 that touches the given 6? Then, how can we place a 5 that touches the given 7? (Strategies 3, 6)

47. How can we fill the empty hexagons that touch the given 3's around the edge of the puzzle? (Strategies 2, 3)

48. How can we fill the empty hexagons that touch the given 5? Later, how can we place a 5 that touches the given 6? (Strategies 3, 6, 8)

49. Where can we place 1's in the top row? Then, how can we fill the empty hexagons that touch the highlighted 6? (Strategies 3, 4, 6)

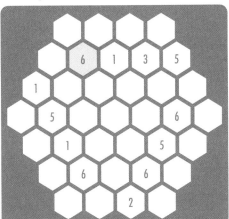

50. The digits 1-6 must appear exactly once each in the empty hexagons that surround the 7. Where can we place a 2 that touches the highlighted 3? Then, where can we place a 2 that touches the highlighted 4? Next, how can we fill the empty hexagons that touch the 7? (Strategies 2, 3, 6)

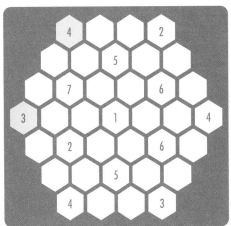

FACTOR CAVE
(Strategies begin on page 92.)

13. Which squares must we shade to contain the 1's? (Strategy 2)

14. Can either 3 clue see 3 squares in the top row? (Strategy 7)

15. Is the square between 2's in the top row shaded or unshaded? (Strategy 7)

16. Which squares must be shaded for the 2 clue in the top-left corner? (Strategies 4, 7)

17. Which squares must be shaded for the 3 clue in the top-left corner? (Strategies 4, 7, 8)

18. How many squares can the 12 and the 6 see in the top row? (Strategy 8)

19. Can the two 6's in the top row see each other? (Strategy 7)

20. Look at the 8's in the corner and on the edges. What is the smallest factor of 8 we can use? Which squares can we mark as unshaded? (Strategies 4, 5)

21. Which squares must be unshaded for the 9? The 6? The 4? (Strategies 4, 5)

22. Which squares must be unshaded for the 12? For the 8 in the bottom-right corner? (Strategy 5)

23. How many squares must the 4 and the 6 in the top row both see? (Strategy 8)

24. Which squares must be unshaded for the 12? Then, is the highlighted square below shaded or unshaded? (Strategies 4, 5, 7)

25. Work from the largest numbers to the smallest. (Strategy 6)

26. Which squares must we shade to contain the 1's? Which other clues have limited options? (Strategies 2, 3)

27. Which squares must be unshaded for the 15? The 20? (Strategies 3, 5, 6)

28. Which numbers in the middle row can see each other? (Strategy 8)

29. Can the 20 and the 15 in the bottom row see each other? (Strategy 8)

30. Which squares must be unshaded for the 16? (Strategies 4, 6)

31. Which squares must be unshaded for the 12? Then, what can we determine using the 4 clues? (Strategies 5, 7)

32. Which squares must be unshaded for the 16? Then, how can we use the 4 clue in the top-right corner? (Strategies 5, 7)

33. Can we satisfy the 4 in the top row if the 3 and the 6 can see each other? (Strategies 7, 8)

34. How can we satisfy the 4 clue in the top-left corner? (Strategies 4, 7)

35. How many of the four squares that surround the highlighted 2 can be unshaded? Which of these four squares must be shaded? (Strategy 7)

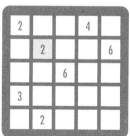

36. Mark squares that must be unshaded for the 25, 20, 15, and 10. Then, is the highlighted square shaded or unshaded? (Strategies 5, 6, 7)

37. Which squares must be unshaded for the 36, 30, and 25? (Strategies 5, 6)

38. Can the 12 and 10 clues in the same row see each other? How many squares can the 8 see in its column? (Strategies 7, 8)

39. Which squares must be unshaded for the 24? Then, can the 16 clue see the highlighted square? (Strategies 3, 5, 7)

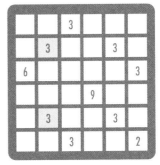

40. What squares must be unshaded for the 20? The 16? The 15? (Strategies 5, 6)

41. How can we complete the highlighted 4 clue? Consider the numbers in the bottom row. (Strategy 7)

42. What squares must be unshaded for each number? Then, what can we determine using the 8? The 10 in the third row? (Strategies 5, 7)

43. Can the 6's in the left column see each other? Consider the 30. (Strategies 5, 7)

44. Can the entire left column be completely unshaded? What square(s) must be unshaded for the 6? Then, how can we complete the highlighted 3 in the second row? What does this tell us about the 3 below it?

45. Which squares must be unshaded for the 35? Can the 7 see the 4 in its row? Can the other 7 see the 20? Then, how can the 21 see 7 squares? (Strategies 3, 7)

46. Which squares must be unshaded for each 24? Then, what factor combinations can we use to satisfy the 24's in the top row? Is there only one combination that works? (Strategies 5, 7, 8)

FACTOR BLOBS
(Strategies begin on page 104.)

19. Which walls can we draw between 3's? (Strategy 3)

20. Which walls can we draw between 7's? Between 4's? (Strategy 3)

21. Which 3 must each 7 be paired with? (Strategies 5, 6, 7)

22. Which blob can contain the 9 in the bottom-left corner? (Strategy 2)

23. Which factor(s) must be grouped with the 6? (Strategies 5, 6)

24. How many factors of 2 must each blob contain? (Strategy 6)

25. Which walls can we draw? Then, which blob is the 26 part of? (Strategies 3, 4)

26. Which walls can we draw between multiples of 5? (Strategies 3, 7)

27. Which walls can we draw? (Strategies 2, 3)

28. How can we pair each 10 with a 7? (Strategy 6)

29. Which blobs can contain 2's? How can we create them without isolating regions? (Strategies 5, 6)

30. Which walls can we draw? (Strategy 3)

31. Which walls can we draw? (Strategy 3)

32. Which multiple of 3 can we connect to the 4? (Strategies 5, 6)

33. Which number must we connect to the 9? To the 3? (Strategy 6)

34. Which walls can we draw? Then, which blob is the bottom-left 6 part of? (Strategies 2, 3, 8)

35. What walls can we draw around the 25's? Then, what blob can contain the 5 in the bottom-right corner? (Strategies 2, 3)

36. Which multiple of 5 must the 50 be connected to? Then, which blob can contain the 4 in the top-right corner? (Strategies 2, 6)

37. Which blob can we complete with the 10 in the bottom-left corner? With the 20 in the bottom-right corner? (Strategy 2)

38. Which 2 can only be grouped with the 11 in the bottom row? Then, what other 2's can we pair with 11's? (Strategy 6)

39. Which blob is the highlighted 3 part of? (Strategy 1)

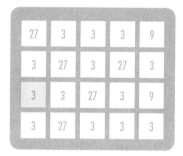

40. What other numbers must be in the same blob as the 10? How can we connect them without creating isolated regions? (Strategies 5, 6, 7)

41. Which walls can we draw between 5's? (Strategy 3)

42. Which pairs of factors give a product of 48? How can we connect them without blocking other pairs? (Strategies 5, 6)

43. After drawing walls, what blob can we make using the 2 in the top-left corner that does not isolate larger numbers? (Strategies 3, 5)

44. How can we pair the 10 in the bottom row with a multiple of 3? Then, which 5 can we connect to the 2 in the bottom-right corner? (Strategies 6, 7)

FRACTION SUMDOKU
(Strategies begin on page 118.)

7. Which two digits must be placed in the 7 cage? Then, what fraction must go in the bottom-right corner? (Strategies 4, 5)

8. How can we place a fraction in the top-right cage without making the sum of the numbers in the 5 cage too large? (Strategy 7)

9. After filling in the center cage, where can the 3 and 1 be placed in the right column? (Strategies 1, 3)

10. Which three digits must be placed in the 11 cage? (Strategy 8)

11. After filling the cages with only one square, what digits must the fractions in the left column and bottom row have? Then, what number must go in the center square? (Strategies 1, 6)

12. How can we fill the 11 cage? (Strategy 8)

13. Where can the 4's be placed? Where can the 1's be placed? (Strategies 1, 3)

14. Which two digits must be placed in the 4 cage? Then, what fraction must go in the top-right corner? (Strategies 4, 5)

15. Which number and fraction must be placed in the 3 cage? (Strategy 9)

16. How can we fill the 4 cage? (Strategy 8)

17. Which two digits must be placed in the 7 cage? Then, what fraction must go in the top-left corner? (Strategies 4, 5)

18. How can we fill the lower 7 cage without making the sum too large? (Strategies 4, 6, 8, 9)

19. After filling the cages with only one square, how can we fill the cages that touch the 1 cage? (Strategies 1, 5)

20. How can we complete the top row? The left column? (Strategies 1, 4, 5, 6)

21. Which two digits must be placed in the 4 cage? Then, how can we complete the 7 cage below it? (Strategies 4, 5)

22. How can we fill the 2 cage in the bottom row without making the sum of the squares in the 6 cage too large? Then, how can we complete the top row? (Strategies 3, 4, 6, 7)

23. How can we fill the 9 cage? The 8 cage? (Strategies 1, 4)

24. How can we fill the bottom-right 3 cage? (Strategy 9)

25. Which two digits must fill the top-right 3 cage? Which two digits must fill the bottom-left 4 cage? Then, where can the 3 be placed in the top row? Where can the 2 be placed in the left column? (Strategies 3, 4, 8)

26. How can we fill the 14 cage? Then, what numbers can fill the 7 cage and the 9 cage? (Strategies 4, 8)

27. After filling in the cages with only one square, where can we place the remaining 5's? (Strategies 1, 3, 10)

28. Which two digits must be placed in the 9 cage? Which three digits must be placed in the 7 cage? Then, what are the denominators of the fractions in the 4 cage? (Strategies 4, 10)

29. After filling in the cages with only one square, which two digits must be placed in the 3 cage? Then, what are the denominators of the fractions in the 6 cage? (Strategies 1, 4, 10)

30. After filling in the cages with only one square, which three digits must be placed in the 6 cage? Then, what are the denominators of the fractions in the top-left 5 cage? (Strategies 1, 4, 10)

31. Which two digits must be placed in the top-left 4 cage? Then, how can we fill the top-right 4 cage? (Strategies 4, 6)

32. How can we place the non-fraction digits in the 2 and 4 cages without making their sums too large? Then, what fractions must fill the 2 cage? (Strategies 6, 7, 8)

33. How can we fill the fractions in all the cages that have one square without repeating digits in a row? Then, how can we fill the 26 cage without making the sum too small? (Strategies 1, 6, 8)

34. There are three ways two fractions can fill the 1 cage without repeating any digits. Which of these can work? Then, what digit fills the top-right corner? What are the possible ways to fill the top-left and bottom-left corners? (Strategies 5, 6, 7)

35. What possible pairs of digits can go in the top-left 6 cage? How can we fill the top-right 5 cage? Then, which digits must be placed in the other cages that have two squares? How can we fill the 9 cage? (Strategies 4, 6, 10)

36. How can we fill the top-left 2 cage without making the sums of the 9 or 6 cages too large? Then, how can we fill the 4 cage? Then, what fraction must go in the bottom-right 2 cage? (Strategies 1, 4, 6, 7)

37. What sets of digits can fill the cages that have 4 squares? (Each has two possibilities.) What digits must be used in the highlighted fractions?

What is the fraction in the 8 cage? How can we fill both fractions in the top-left 12 cage to make its sum 12? How can we fill both fractions in the 20 cage to make its sum 20? (Strategies 4, 6, 10)

38. After filling in the cages with only one square, how can we fill the fraction in the 8 cage? What digits fill the other squares in that cage?

Then, how can we fill the fraction in the 7 cage? What digits fill the other squares in that cage?

Next, how can we fill the 4 cages? (Strategies 1, 5, 6, 8, 10)

SUM SQUARES
(Strategies begin on page 132.)

19. Which pairs of numbers can complete the middle row? The middle column? (Strategies 3, 4, 5)

20. Which pairs of numbers can complete the bottom row? Then, how can we complete the top row? (Strategies 3, 7)

21. Which pairs of numbers can complete the top row? The right column? (Strategies 3, 4, 5)

22. Which pairs of numbers can complete the left column? The middle row? (Strategies 3, 4, 5)

23. After completing the middle column and top row, what pair of numbers can complete the right column? How can we place them? (Strategies 2, 5)

24. After completing the bottom row and middle column, which pairs of numbers can complete the right column? The top row? (Strategies 2, 3, 4)

25. Which pair of numbers can complete the middle row? Where can we place them? (Strategies 3, 6)

26. After completing the top row, which pair of numbers can complete the right column? Then, which pair of numbers can complete the left column? (Strategies 2, 3, 5)

27. Which pair of numbers can complete the middle row? Where can we place them? (Strategies 3, 5, 6)

28. Which pair of numbers can complete the middle row? Where can we place them? (Strategies 3, 5, 6)

29. Which pair of numbers can complete the left column? The middle column? (Strategies 3, 7)

30. After completing the right column, which pair of numbers can complete the middle row? Where can we place them? (Strategies 2, 3, 5, 6)

31. Which pairs of numbers can complete the left column? The middle row? (Strategies 3, 4)

32. Which pairs of numbers can complete the right column? Where can we place them? (Strategies 3, 5, 6)

33. Which three numbers sum to 24? Which of these numbers can fill the bottom-left corner? (Strategies 5, 6)

34. Which pair of numbers can complete the middle column? Given those choices, how can we complete the middle row? (Strategies 3, 7, 8)

35. Which pairs of numbers can complete the middle row? The middle column? What does this give us for the middle square? (Strategies 3, 4)

36. Both the top row and left column need pairs of numbers that sum to 14. Which remaining numbers must go in the other two empty squares? How can we complete the middle column? (Strategies 3, 4, 6)

37. Which three numbers go in the top row? Which of these can we place in the middle column? (Strategies 5, 6, 7)

38. Which three numbers give a sum of 24? Then, which three numbers give a sum of -15? (Strategy 5)

39. Which pairs of numbers can complete the bottom row? The left column? What two possibilities does this give us for the bottom-left square? (Strategies 3, 4)

40. Which pairs of numbers can complete the top row? The left column? What two possibilities does this give us for the top-left square? (Strategies 3, 4)

41. After completing the top row, which pairs of numbers complete each column? Which of the pairs in the left column lets us complete the other columns? (Strategies 2, 3, 7)

42. Which groups of numbers can complete the right column? How can the middle row help us determine which of these sets works? Then, what pair of numbers can complete the middle column?
(Strategies 3, 4, 6, 7)

PAINT THE TOWN
(Strategies begin on page 146.)

19. What do the clues below tell us about the highlighted houses? (Strategy 5)

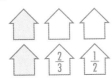

20. Which house of the six houses must be shaded below? (Strategy 5)

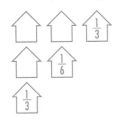

21. What do the clues below tell us about the highlighted houses? (Strategy 5)

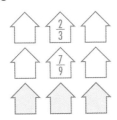

22. What do the clues below tell us about the highlighted houses? (Strategy 5)

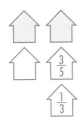

23. Start at the 0 and work up and then around. (Strategies 2, 3)

24. What do the clues below tell us about the highlighted houses? (Strategy 5)

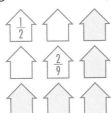

25. What do the clues below tell us about the highlighted houses? (Strategy 5)

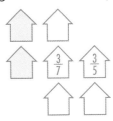

26. What do the clues below tell us about the highlighted houses? (Strategies 2, 3)

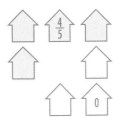

27. What do the clues below tell us about the highlighted house? (Strategy 5)

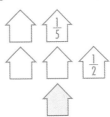

28. What do the clues below tell us about the highlighted house? (Strategy 5)

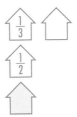

29. What do the clues below tell us about the highlighted houses? (Strategies 2, 5)

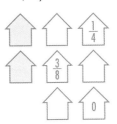

30. What do the clues below tell us about the highlighted houses? (Strategies 2, 3)

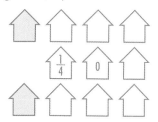

31. What do the clues below tell us about the highlighted houses? (Strategy 5)

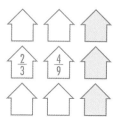

32. What do the clues below tell us about the highlighted house? (Strategy 5)

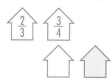

33. Start at the 1 in the top-right corner and work clockwise. (Strategies 2, 3)

34. Which houses must be shaded or unshaded below? (Strategies 2, 3)

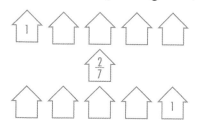

35. What do the clues below tell us about the highlighted houses? (Strategy 5)

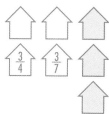

36. What do the clues below tell us about the highlighted houses? (Strategy 5)

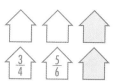

37. What do the clues below tell us about the highlighted houses? (Strategy 5)

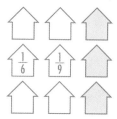

38. Start in the top-left corner and work from left to right, completing each column one at a time. (Strategies 2, 3, 5)

39. Which houses must be shaded or unshaded below? (Strategies 2, 3)

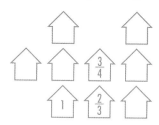

40. Which houses must be shaded or unshaded below? (Strategies 2, 3)

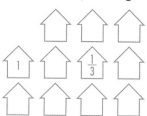

41. What do the clues below tell us about the highlighted houses? (Strategy 5)

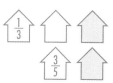

42. How could we use the $\frac{1}{2}$ and the adjacent $\frac{2}{3}$ in each corner to shade some houses? (Strategy 5)

43. What do the clues below tell us about the highlighted house? How can we apply the same strategy elsewhere? (Strategy 5)

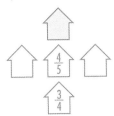

44. What do the clues below tell us about the highlighted houses? (Strategy 5)

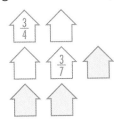

45. What do the clues below tell us about the highlighted houses? (Strategy 5)

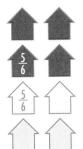

46. What do the clues below tell us about the highlighted house? Then, how can we use the $\frac{1}{9}$ clue? (Strategies 2, 5, 6)

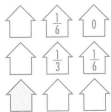

47. Start at the bottom and work up. In each row, compare the fraction in the first column to the fraction in the second column. What do the clues tell us about the highlighted houses in the third column? (Strategy 5)

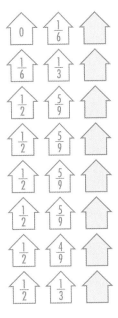

DECIMAL NUMBERCROSS
(Strategies begin on page 160.)

23. Where can we place $\frac{3,553}{100}$?
(Strategy 3)

24. Where can we place $\frac{9,999}{10,000}$?
(Strategy 3)

25. Where can we place $\frac{2,002}{10}$?
(Strategy 3)

26. Where can we place $\frac{1,055}{10}$?
(Strategy 3)

27. What number can we place in the middle row? (Strategy 4)

28. Where can we place the two numbers that are greater than 1?
(Strategies 3, 6)

29. Where can we place $\frac{2,006}{1,000}$? (Strategy 3)

30. Where can we place the largest numbers? (Strategies 3, 5)

31. Where can we place $\frac{444}{1,000}$? (Strategy 5)

32. What digits of the largest numbers can we place? (Strategies 3, 6)

33. What digit can we place in the highlighted square? (Strategy 6)

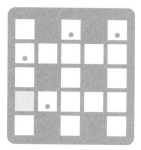

34. There are three places where we can place $\frac{43}{10}$ and $\frac{45}{10}$. Can we place any of those digits? (Strategies 5, 6)

35. Where can we place $\frac{306}{100}$?
(Strategies 3, 5)

36. Which number can we place in the middle row? (Strategy 4)

37. Where can we place $\frac{104}{10}$?
(Strategy 3)

38. Where can we place $\frac{1,234}{100}$?
(Strategy 3)

39. Where can we place $\frac{123}{10,000}$?
(Strategies 3, 5)

40. Where can we place the decimal points in the grid so that there is exactly one decimal point in each number? (Strategy 7)

41. Where can we place the decimal points? (Strategy 7)

42. Which digit is not repeated? Where can we place it? (Strategies 4, 5)

43. There are three numbers that require at least 4 squares. How can we place the decimal points for these three numbers? (Strategies 5, 7)

44. Which squares cannot contain a decimal point? (Strategy 7)

SOLUTIONS

DOT PUZZLES SOLUTIONS

Page 7

1.

2.

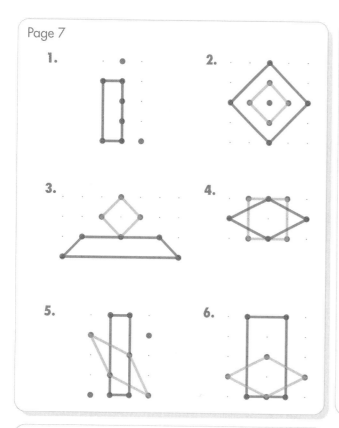

3.

4.

5.

6.

Page 8

7.

8.

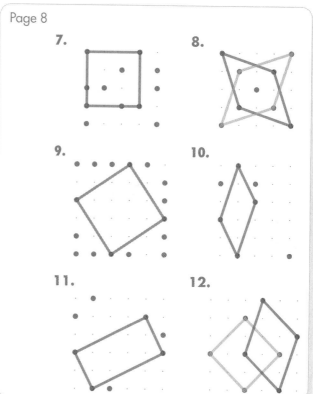

9.

10.

11.

12.

Page 9

13.

14.

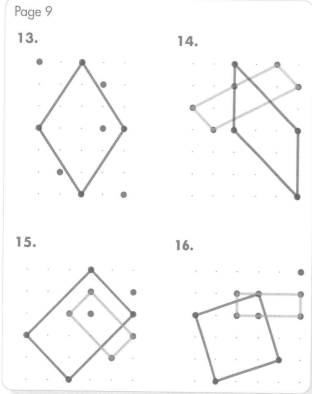

15.

16.

Page 10

17.

18.

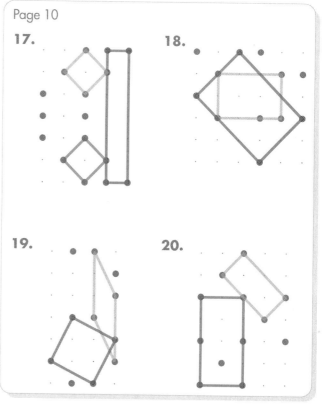

19.

20.

Page 11

21.

22.

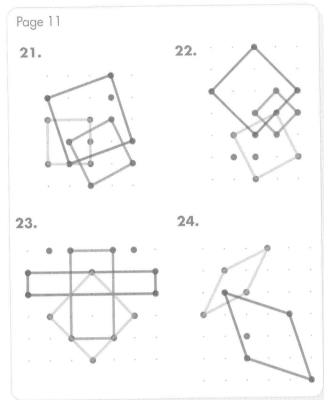

23.

24.

Page 12

25.

26.

27.

28.

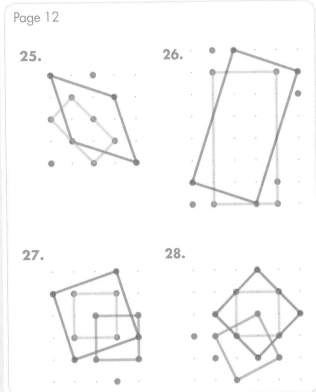

Page 13

29.

30.

31.

32.

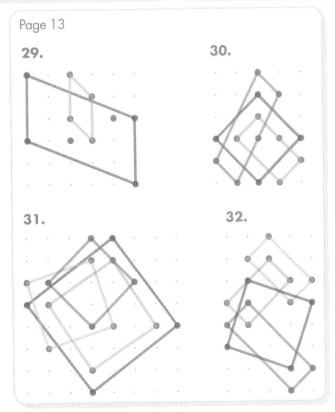

33. **34.**

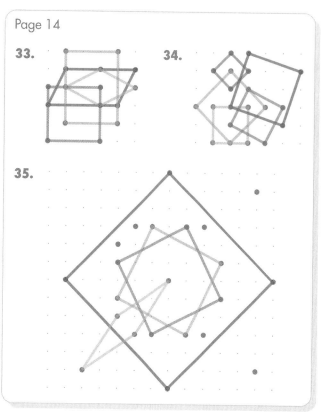

35.

36. **37.**

38.

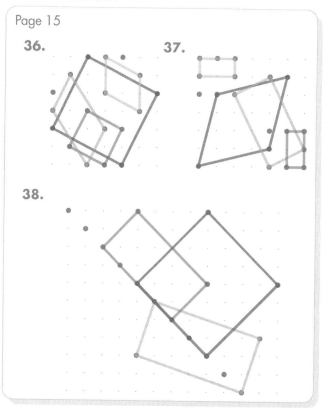

SPIRAL GALAXIES SOLUTIONS

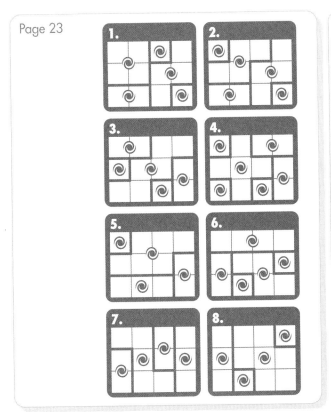

Page 23

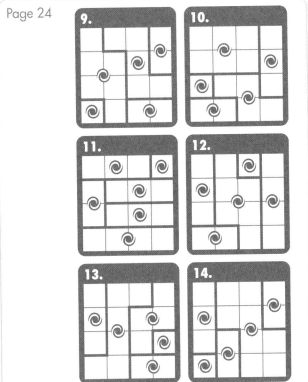

Page 24

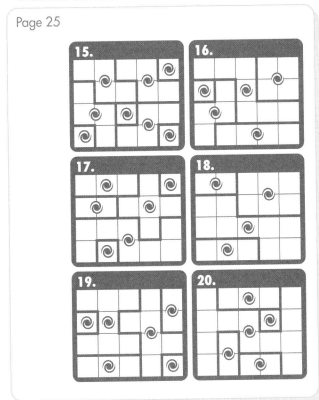

Page 25

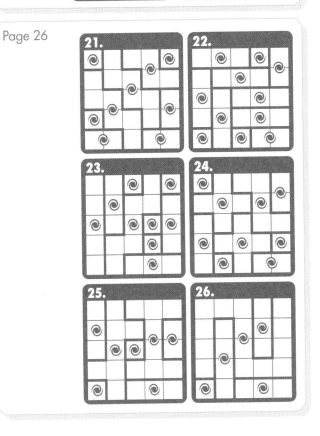

Page 26

Beast Academy Puzzles 4 - Solutions

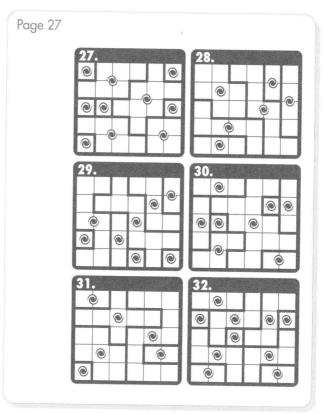

Page 27

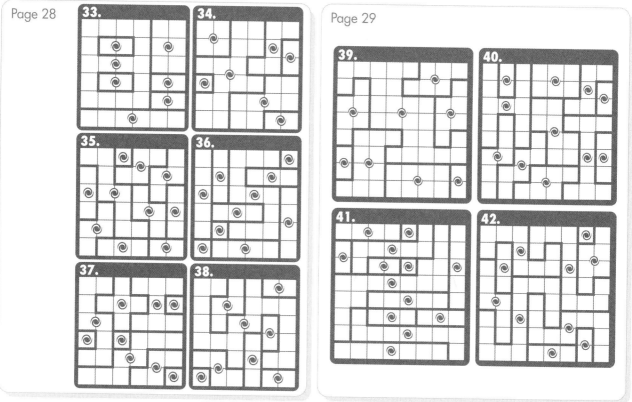

Page 28

Page 29

PRODUCT PLACEMENT SOLUTIONS

Page 35

1.
```
    8 5
  ×   2
  ─────
    1 0
+ 1 6 0
─────────
  1 7 0
```

2.
```
    2 8
  ×   8
  ─────
    6 4
+ 1 6 0
─────────
  2 2 4
```

3.
```
    7 9
  ×   4
  ─────
    3 6
+ 2 8 0
─────────
  3 1 6
```

4.
```
    6 4
  ×   4
  ─────
    1 6
+ 2 4 0
─────────
  2 5 6
```

5.
```
    2 6
  ×   7
  ─────
    4 2
+ 1 4 0
─────────
  1 8 2
```

6.
```
    7 9
  ×   7
  ─────
    6 3
+ 4 9 0
─────────
  5 5 3
```

Page 36

7.
```
    8 5
  ×   6
  ─────
    3 0
+ 4 8 0
─────────
  5 1 0
```

8.
```
    5 4
  ×   7
  ─────
    2 8
+ 3 5 0
─────────
  3 7 8
```

9.
```
    3 7
  ×   6
  ─────
    4 2
+ 1 8 0
─────────
  2 2 2
```

10.
```
    2 4
  ×   7
  ─────
    2 8
+ 1 4 0
─────────
  1 6 8
```

11.
```
    9 8
  ×   6
  ─────
    4 8
+ 5 4 0
─────────
  5 8 8
```

12.
```
    5 7
  ×   3
  ─────
    2 1
+ 1 5 0
─────────
  1 7 1
```

Page 37

13.
```
    6 5
  ×   6
  ─────
    3 0
+ 3 6 0
─────────
  3 9 0
```

14.
```
    7 8
  ×   9
  ─────
    7 2
+ 6 3 0
─────────
  7 0 2
```

15.
```
    4 6
  ×   6
  ─────
    3 6
+ 2 4 0
─────────
  2 7 6
```

16.
```
    5 9
  ×   8
  ─────
    7 2
+ 4 0 0
─────────
  4 7 2
```

17.
```
    4 7
  ×   5
  ─────
    3 5
+ 2 0 0
─────────
  2 3 5
```

18.
```
    9 4
  ×   9
  ─────
    3 6
+ 8 1 0
─────────
  8 4 6
```

Page 38

19.
```
    4 9
  ×   7
  ─────
    6 3
+ 2 8 0
─────────
  3 4 3
```

20.
```
    3 7
  ×   3
  ─────
    2 1
+   9 0
─────────
  1 1 1
```

21.
```
    9 6 3
  ×     7
  ───────
      2 1
    4 2 0
+ 6 3 0 0
───────────
  6 7 4 1
```

22.
```
    7 8 5
  ×     4
  ───────
      2 0
    3 2 0
+ 2 8 0 0
───────────
  3 1 4 0
```

23.
```
    1 1 7
  ×     7
  ───────
      4 9
      7 0
+   7 0 0
───────────
    8 1 9
```

24.
```
    7 3 6
  ×     6
  ───────
      3 6
    1 8 0
+ 4 2 0 0
───────────
  4 4 1 6
```

25.
```
      6 3 6
    ×     8
    ─────────
      4 8
    2 4 0
  + 4 8 0 0
  ─────────
    5 0 8 8
```

26.
```
      8 1 2
    ×     8
    ─────────
      1 6
      8 0
  + 6 4 0 0
  ─────────
    6 4 9 6
```

27.
```
      1 1 4
    ×     9
    ─────────
      3 6
      9 0
  +   9 0 0
  ─────────
    1 0 2 6
```

28.
```
      3 4 6
    ×     9
    ─────────
      5 4
    3 6 0
  + 2 7 0 0
  ─────────
    3 1 1 4
```

29.
```
      2 6 6
    ×     4
    ─────────
      2 4
    2 4 0
  +   8 0 0
  ─────────
    1 0 6 4
```

30.
```
      9 1 2
    ×     5
    ─────────
      1 0
      5 0
  + 4 5 0 0
  ─────────
    4 5 6 0
```

31.
```
      1 5 5
    ×     9
    ─────────
      4 5
    4 5 0
  +   9 0 0
  ─────────
    1 3 9 5
```

32.
```
      3 5 4
    ×     9
    ─────────
      3 6
    4 5 0
  + 2 7 0 0
  ─────────
    3 1 8 6
```

33.
```
      4 2 9
    ×     7
    ─────────
      6 3
    1 4 0
  + 2 8 0 0
  ─────────
    3 0 0 3
```

34.
```
      1 1 9
    ×     9
    ─────────
      8 1
      9 0
  +   9 0 0
  ─────────
    1 0 7 1
```

35.
```
      8 9 2
    ×     9
    ─────────
      1 8
      8 1 0
  + 7 2 0 0
  ─────────
    8 0 2 8
```

36.
```
      6 7 8
    ×     7
    ─────────
      5 6
      4 9 0
  + 4 2 0 0
  ─────────
    4 7 4 6
```

37.
```
      5 4 6 7
    ×       3
    ───────────
          2 1
        1 8 0
      1 2 0 0
  + 1 5 0 0 0
  ───────────
    1 6 4 0 1
```

38.
```
      5 1 1 7
    ×       8
    ───────────
          5 6
          8 0
        8 0 0
  + 4 0 0 0 0
  ───────────
    4 0 9 3 6
```

39.
```
      1 3 4 9
    ×       7
    ───────────
          6 3
        2 8 0
      2 1 0 0
  +   7 0 0 0
  ───────────
      9 4 4 3
```

40.
```
      4 3 8 9
    ×       9
    ───────────
          8 1
        7 2 0
      2 7 0 0
  + 3 6 0 0 0
  ───────────
    3 9 5 0 1
```

Beast Academy Puzzles 4 - Solutions

PYRAMID DESCENT SOLUTIONS

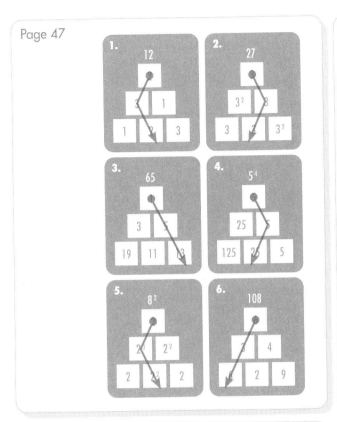

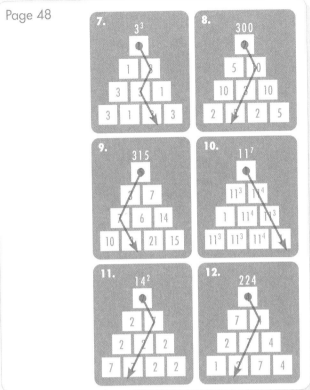

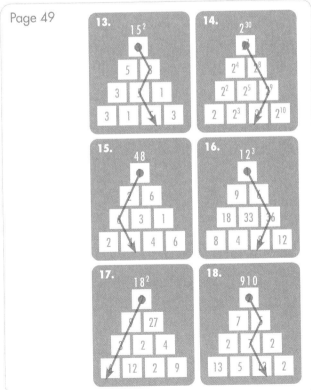

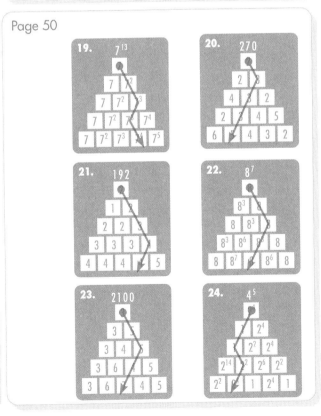

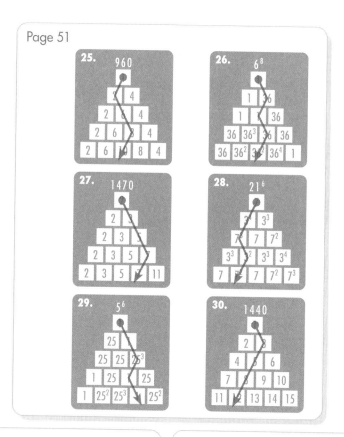

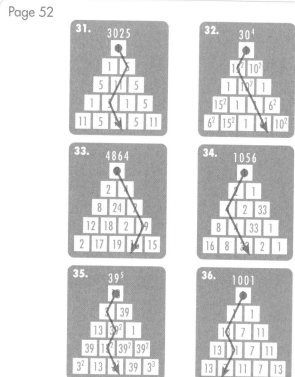

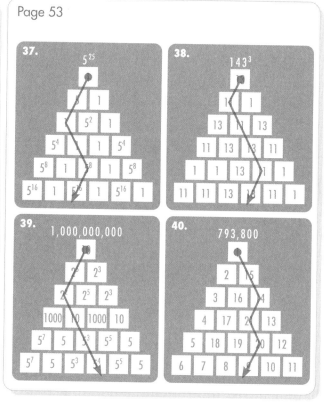

DUTCH LOOP SOLUTIONS

Page 57

Page 58

Page 59

Page 60

Page 61

Page 62

Page 63

HIVE SOLUTIONS

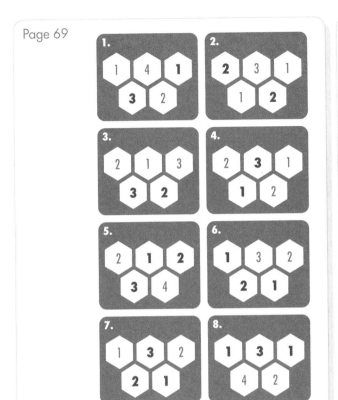

Page 69

1.
1 4 1
3 2

2.
2 3 1
1 2

3.
2 1 3
3 2

4.
2 3 1
1 2

5.
2 1 2
3 4

6.
1 3 2
2 1

7.
1 3 2
2 1

8.
1 3 1
4 2

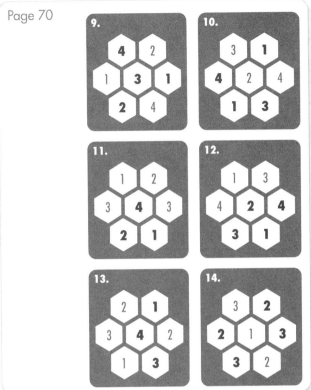

Page 70

9.
4 2
1 3 1
2 4

10.
3 1
4 2 4
1 3

11.
1 2
3 4 3
2 1

12.
1 3
4 2 4
3 1

13.
2 1
3 4 2
1 3

14.
3 2
2 1 3
3 2

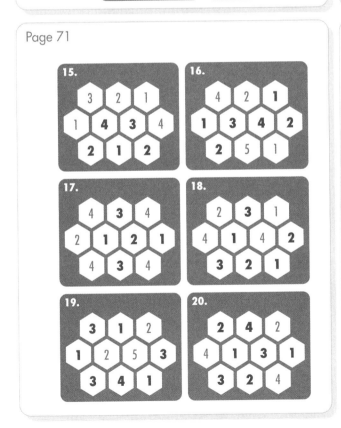

Page 71

15.
3 2 1
1 4 3 4
2 1 2

16.
4 2 1
1 3 4 2
2 5 1

17.
4 3 4
2 1 2 1
4 3 4

18.
2 3 1
4 1 4 2
3 2 1

19.
3 1 2
1 2 5 3
3 4 1

20.
2 4 2
4 1 3 1
3 2 4

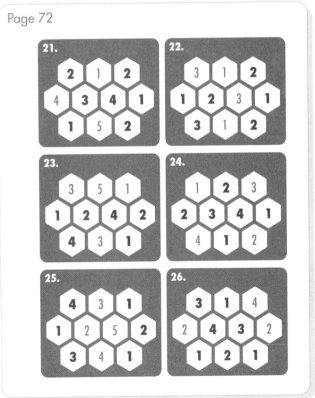

Page 72

21.
2 1 2
4 3 4 1
1 5 2

22.
3 1 2
1 2 3 1
3 1 2

23.
3 5 1
1 2 4 2
4 3 1

24.
1 2 3
2 3 4 1
4 1 2

25.
4 3 1
1 2 5 2
3 4 1

26.
3 1 4
2 4 3 2
1 2 1

Page 73

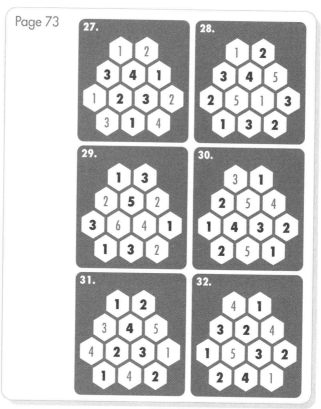

Page 74

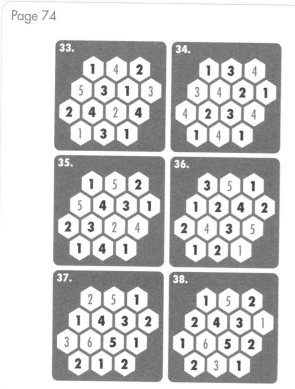

Page 75

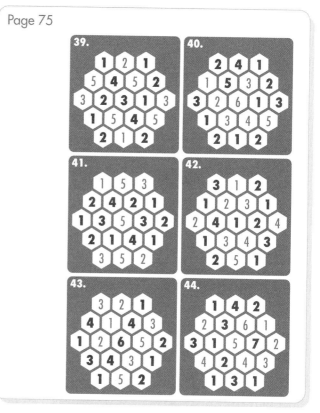

Page 76

45.
46.

47.
48.

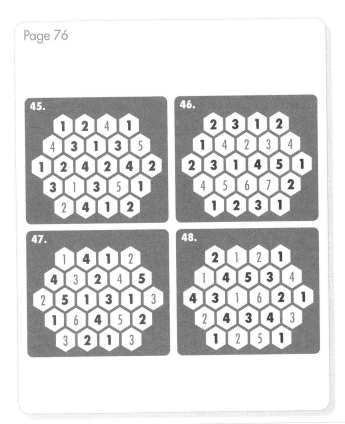

Page 77

49.

50.

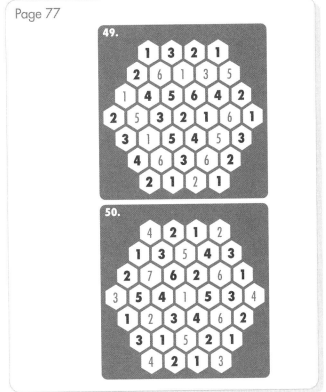

FACTOR CAVE SOLUTIONS

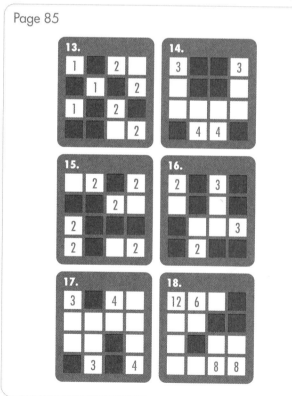

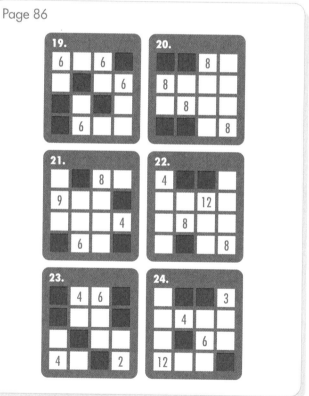

206

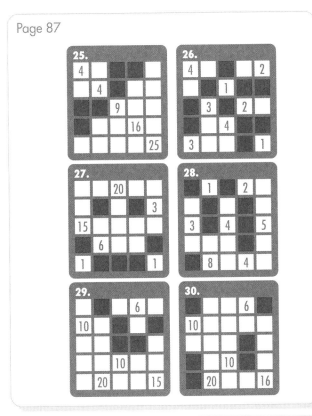

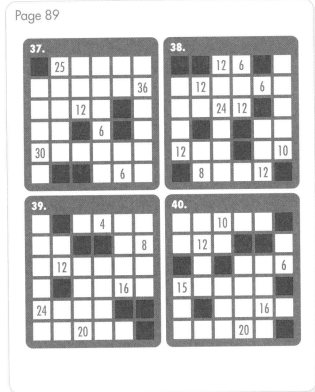

Page 90

Page 91

FACTOR BLOBS SOLUTIONS

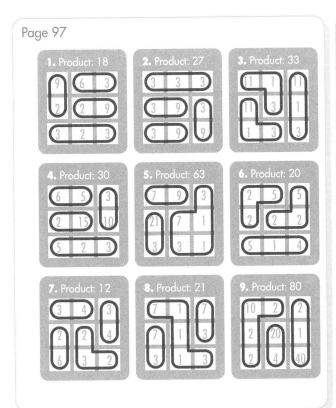

Page 97

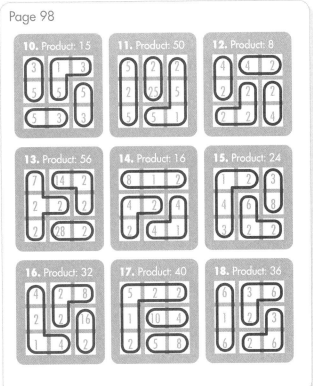

Page 98

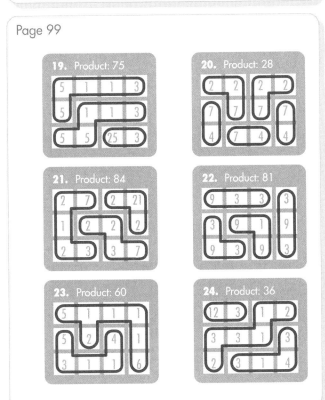

Page 99

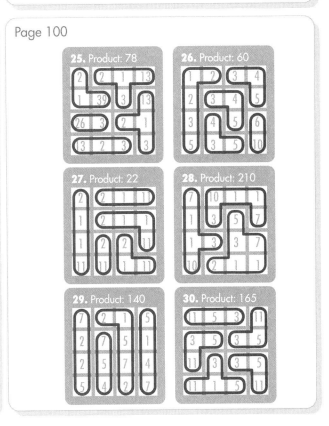

Page 100

Page 101

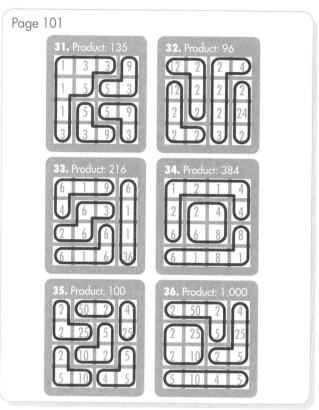

31. Product: 135 **32.** Product: 96
33. Product: 216 **34.** Product: 384
35. Product: 100 **36.** Product: 1,000

Page 102

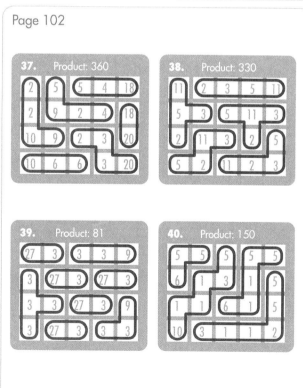

37. Product: 360 **38.** Product: 330
39. Product: 81 **40.** Product: 150

Page 103

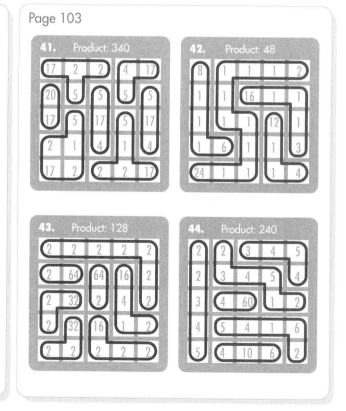

41. Product: 340 **42.** Product: 48
43. Product: 128 **44.** Product: 240

FRACTION SUMDOKU SOLUTIONS

Page 109

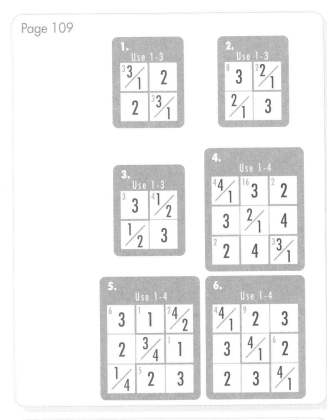

Page 110

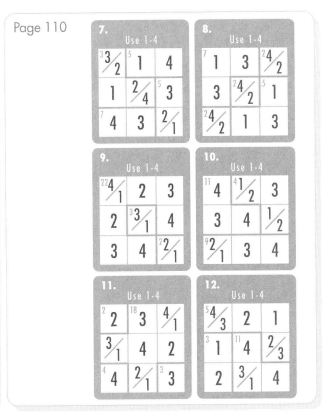

Page 111

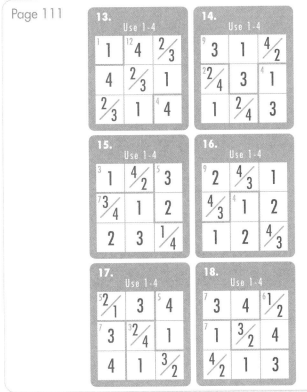

Page 112

Page 113

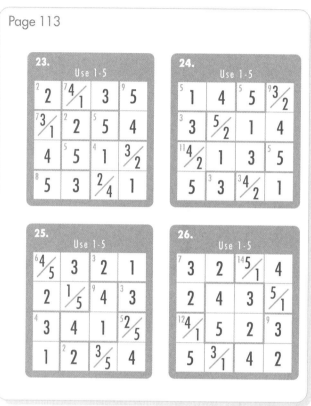

Page 114

Page 115

35.

Use 1-6

⁶2	4	⁹5/6	⁵1	3
⁵3	2	4	5/6	1
²⁰4	¹⁵5	1	3	2/6
5/1	6	3	⁶2	⁹4
6	3/1	2	4	5

36.

Use 1-6

²6/3	⁹4	2	⁶5	1
5	6/2	1	4	3
4	1	²6/3	2	5
2	5	4	¹⁴3/1	6
⁴1	3	5	6	²⁴4/2

37.

Use 1-6

¹²2	4/5	³3	¹³1	²⁰6
3	¹²6	⁸1	5	4/2
1	3	6/2	4	5
6/5	2	4	3	1
4	1	⁵5	6/2	3

38.

Use 1-7

¹1	⁴7/5	⁸2	4	6/3
3/5	2	¹⁴7/4	¹²6	1
¹⁷7	3/4	6	1/2	5
4	6	5/1	⁴3	2/7
⁷6/2	1	3	5/7	⁴4

SUM SQUARES SOLUTIONS

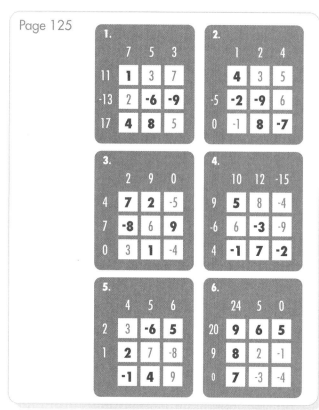

Page 125

1.

	7	5	3
11	**1**	3	7
-13	2	**-6**	**-9**
17	**4**	**8**	5

2.

	1	2	4
	4	3	5
-5	**-2**	**-9**	6
0	-1	**8**	**-7**

3.

	2	9	0
4	**7**	**2**	-5
7	**-8**	6	**9**
0	3	**1**	-4

4.

	10	12	-15
9	**5**	8	-4
-6	6	**-3**	-9
4	**-1**	**7**	**-2**

5.

	4	5	6
2	3	**-6**	**5**
1	**2**	7	-8
	-1	**4**	9

6.

	24	5	0
20	**9**	**6**	**5**
9	**8**	2	-1
0	**7**	-3	-4

Page 126

7.

	18	-5	18
9	**6**	**2**	1
13	**7**	-3	**9**
9	5	**-4**	8

8.

	1	11	21
10	**4**	**-3**	9
11	-2	**6**	**7**
12	-1	8	**5**

9.

	5	2	2
6	2	**1**	3
6	**9**	**-7**	**4**
-3	-6	**8**	-5

10.

	-1	0	4
2	-6	**5**	3
-2	**4**	2	**-8**
3	1	**-7**	**9**

11.

	6	-7	
4	**1**	-6	9
8	7	**-4**	**5**
9	**-2**	**3**	8

12.

		1	3
2	**-4**	**-3**	9
4	5	**-2**	**1**
7	8	6	**-7**

Page 127

13.

	1	1	3
3	**7**	**-5**	1
3	**-9**	**4**	**8**
-1	3	**2**	-6

14.

	15	15	-15
5	**4**	**8**	**-7**
5	**9**	1	-5
5	**2**	6	**-3**

15.

	-1	15	-9
3	**-2**	9	**-4**
15	**7**	**5**	3
-13	**-6**	1	**-8**

16.

	-2	4	13
3	**-6**	1	**8**
14	**7**	-2	**9**
-2	-3	**5**	-4

17.

	5	5	5
6	**8**	**-3**	1
5	**4**	**6**	**-5**
4	-7	**2**	9

18.

	1	3	5
5	**7**	-5	3
3	**-4**	-1	**8**
1	**-2**	**9**	**-6**

Page 128

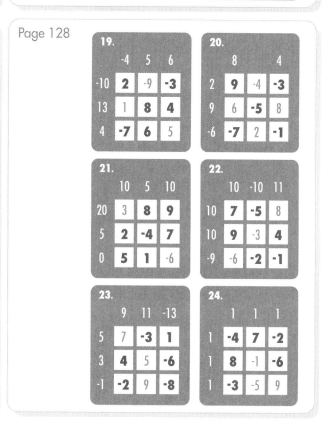

19.

	-4	5	6
-10	**2**	-9	**-3**
13	1	**8**	**4**
4	**-7**	**6**	5

20.

	8		4
2	**9**	-4	**-3**
9	6	**-5**	8
-6	**-7**	2	**-1**

21.

	10	5	10
20	3	**8**	**9**
5	**2**	**-4**	**7**
0	**5**	**1**	-6

22.

	10	-10	11
10	**7**	**-5**	8
10	**9**	-3	**4**
-9	-6	**-2**	**-1**

23.

	9	11	-13
5	7	**-3**	**1**
3	**4**	**5**	-6
-1	**-2**	9	**-8**

24.

	1	1	1
1	**-4**	**7**	**-2**
1	**8**	-1	**-6**
1	**-3**	-5	**9**

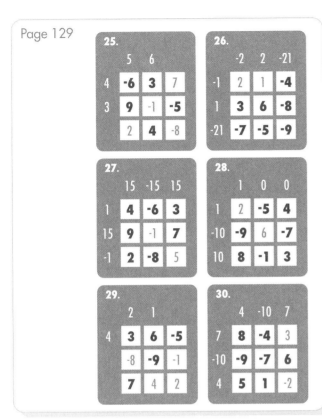

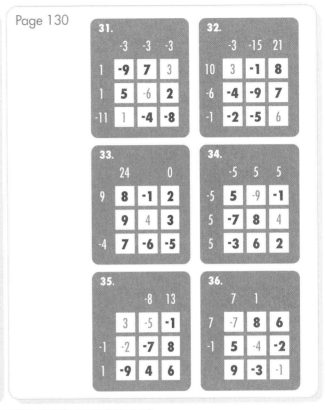

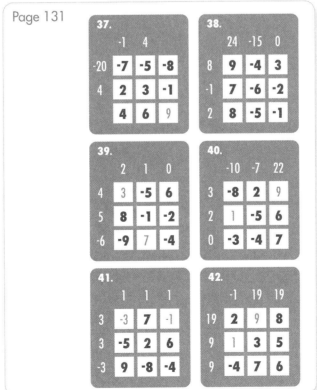

PAINT THE TOWN SOLUTIONS

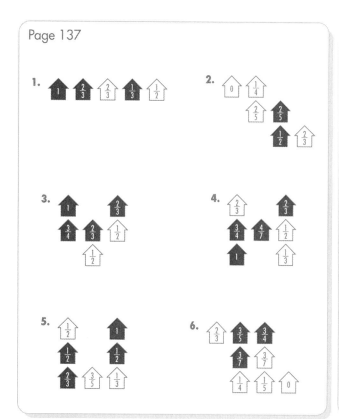

Page 137

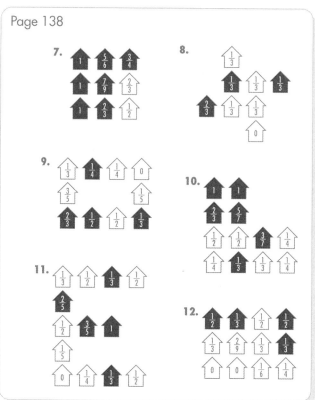

Page 138

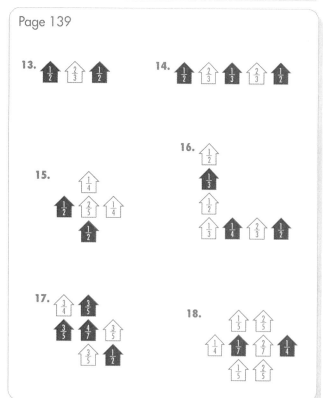

Page 139

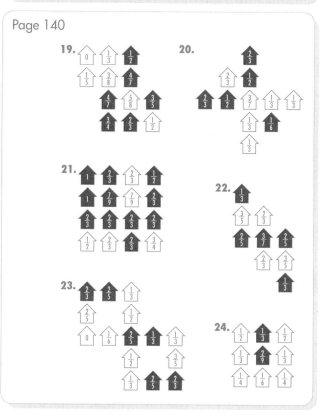

Page 140

Beast Academy Puzzles 4 - Solutions

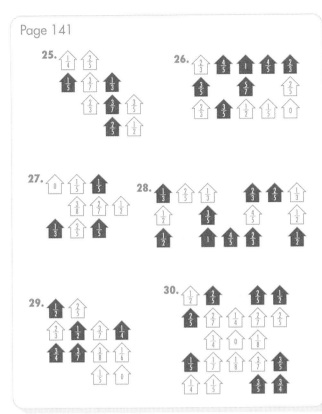

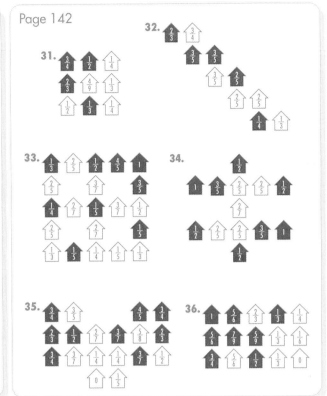

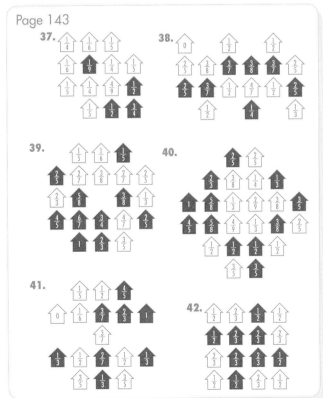

Page 144

43.

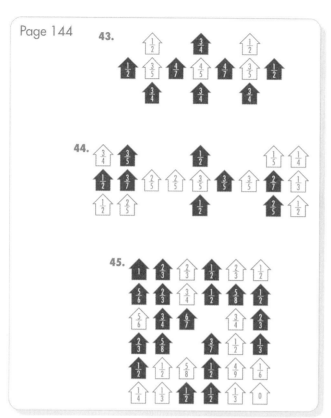

44.

45.

Page 145

46.

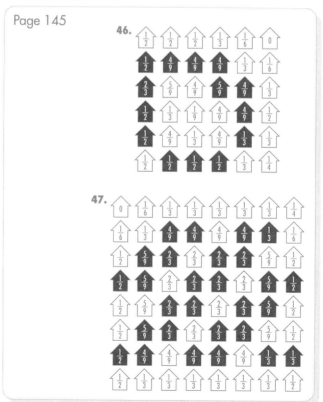

47.

DECIMAL NUMBERCROSS SOLUTIONS

Page 151

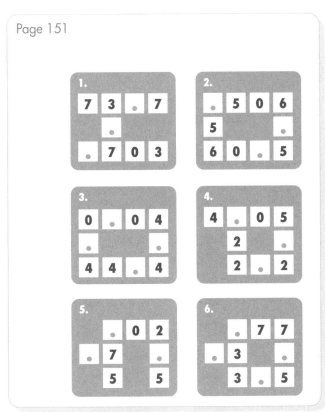

Page 152

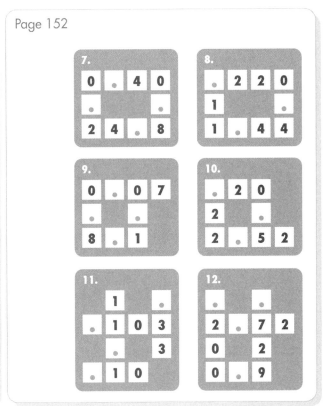

Page 153

Page 154

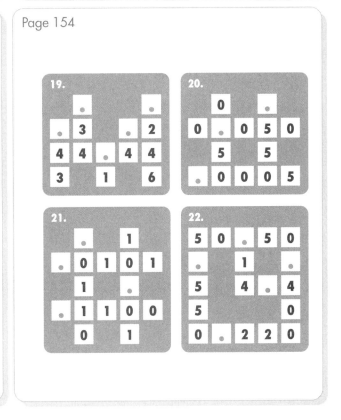

Page 155

Page 156

Page 157

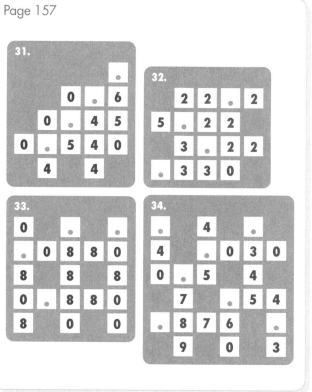

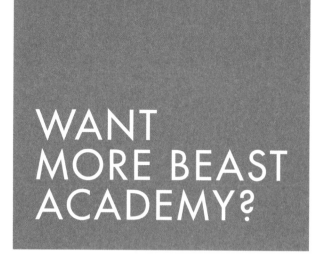

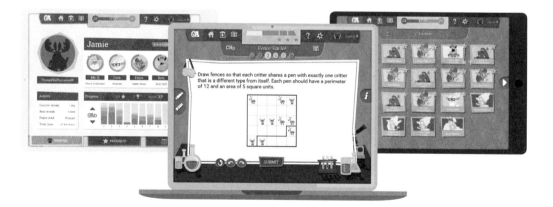